BEES, HONEY AND THE HIVE

Marie-Louise von Franz, Honorary Patron

Studies in Jungian Psychology by Jungian Analysts

Daryl Sharp, General Editor

BEES, HONEY AND THE HIVE

Circumambulating the Centre

A Jungian Exploration of the Symbolism and Psychology

FRITH LUTON

INNER CITY BOOKS

Library and Archives Canada Cataloguing in Publication

Luton, Frith
Bees, honey and the hive: circumambulating the centre: a Jungian exploration of the symbolism and psychology / Frith Luton

(Studies in Jungian psychology by Jungian analysts; 131)
Includes bibliographical references and index
ISBN 978-1-894574-32-7

1. **Bees—Symbolic aspects. 2. Bee culture.**
2. **Jungian psychology. I. Title.**

II. Series: Studies in Jungian psychology by Jungian analysts ; 131

BF175.5.S95L87 2010 302.2'223 C2010-905961-1

INNER CITY BOOKS
Box 1271, Station Q, Toronto, ON M4T 2P4, Canada.
Telephone (416) 927-0355 / Fax (416) 924-1814
Toll-free (Canada and U.S.): Tel. 1-888-927-0355 / Fax 1-888-924-1814
Web site: www.innercitybooks.net /
Email: booksales@innercitybooks.net

Honorary Patron: Marie-Louise von Franz.
Publisher and General Editor: Daryl Sharp.
Senior Editor: Victoria B. Cowan.
Office Manager: Scott Milligen.

INNER CITY BOOKS was founded in 1980 to promote the understanding and practical application of the work of C. G. Jung.

Cover Image: Bee gathering nectar; photographer unknown.

Printed and bound in Canada by Thistle Printing Limited

CONTENTS

List of Figures 7

Introduction 9
Biographical 12

1 The Bee in Nature and the Natural Sciences 17
The Queen Bee and the Hive 18
Nectar, Honey and Tapas 24
Worker bees 26
Spring and Kundalini rising 26
Summer: Navigation and Dance Language 27
Wintering: The *Nigredo* 31
Swarming: Chaos and Interregnum—A Move Towards New Order 32
Colony Collapse Disorder 36
Beekeepers 38
The Well-regulated Hive 39
Ultra-violet and Infra-red: the Art of Karl von Frisch 41

2 Bee Symbolism in Myth, Legend and Story 46
The Bee Goddess and the Great Mother 47
Feminine Potency and Fertility 47
The Birth of Zeus: Sky and Cave, Light and Honey 55
Dionysos 56
Aristaeus: God of Beekeeping and Bull-born Bees 57
The Bee and the Muses: Inspiration and Divination 64
Delphi and Divination: The *Omphalos* as World Navel and Oracular Centre 65
The Queen and 'King' Bee as Numinous Centre 68
Divine Inspiration: Oracle and Muse 71
Trophonius: the Rainmaker and the Swarm 71
Mercurius/Hermes and the Bee Maidens 73
Creation Myths and Folk Stories 80

3 Bee Symbolism and the Religious 86
Honey and the Religious 87
Ancient Egypt: Re's Tears and the 'Opening of the Mouth' Ceremony 91
Judaism: Wisdom, Sweetness, Samson and Lion's Honey 95

Christianity: Bees, the Hive, the Honey and the Holy Mother 96
Cathedral Builders: Bees and Freemasons 102
Islam: 'The Bee' Sura 104
Hinduism: The Blue Bee and Kama the God of Love 105
Buddhism: The Bee and Buddha Nature 109

4 The Bee, the Honey and the Hive: A Psychological Perspective 112
The Opposites of Love and War 113
Wounding and Healing: Cupid's Arrows, the Bee's Sting and Honey 119
The Transformed Berserker 121

5 Alchemy and the Individuation Process 130
The Alchemical Stages of Making the *Lapis* 131
The *Lapis* and *Multiplicatio* 133
Individuation: The *Opus* 134
Union: Conjunction 134
The *Hierosgamos* 137
The Queen Bee's Nuptial Flight 137
Renewal of the Life of the Hive 139
The Alchemical *Hierosgamos,* the Divine Water and the Gold 141
Deep Eros and the 'Mystery of the Queen' 144
Multiplicatio and Flowering 146
Centring: Circumambulating the Centre 151
Mandala Symbolism: Rotation and *Circulatio* 152
Dorn: The *Caelum* and Three Stages of Illumination or *Coniunctio* 155
Eros and *Multiplicatio:* The Yellowing or *Citrinitas* 162
A Dream of Bees, Honey and the Hive 166

6 Conclusion 170
Reflections 173

Acknowledgements 176

Bibliography 178

Index 191

See final pages for other Inner City titles

List of Figures

1. Worker, drone, queen bee 17
2. Karl von Frisch; bees' circle dance and waggle dance 29
3. Indication of direction by tail-wagging dance 30
4. Some methods of taking a swarm 35
5. Goddess of regeneration: bee-form with antennae 47
6. Gold pendant of two bees 48
7. Bee-headed goddess medallion 49
8. The Goddess Spes with sheaves and beehive 52
9. *Bugonia:* 'bull-born' bees 54
10. Aristaeus 61
11. *Omphalos* stone with net of bees 67
12. Egyptian bee hieroglyphics 70
13. Trophonius with beehive, fruits and flowers 71
14. The Egyptian sun god's birth from a lotus flower 81
15. Venus, the Subject of the Wise 85
16. The Egyptian Goddess Nut, painted ceiling relief 92
17. Mary and Christ child painted on beehive 97
18. Galician Madonna and child 98
19. 'I turn everything to honey' 99
20. 'The rose gives honey to the bee' 102
21. Alchemical image of bees and beehive 103
22. The Indian god Kama riding a bee 106
23. Post-Mesolithic rock art of honey-hunters 111
24. Drawings of bee anatomy by Jan Swammerdam 120
25. A European brown bear drowning bees 126
26. Hermaphrodite 136
27. A spiritual pilgrim glimpses the transcendent world 160
28. *Multiplicatio* image, from *Alchymie* (1606) 162
29. *Multiplicatio* image, from *Cabala* (1616) 164
30. Bees circumambulating the hive 176
31. Danish woodcut showing hives and beekeeping equipment 177

Life is the flower for which love is the honey.
—Victor Hugo.

The life of the bee is like a magic well: the more you draw from it, the more there is to draw.
—Karl von Frisch.

In the honey, the 'sweetness of earths', we can easily recognise the balsam of life that permeates all living, budding, and growing things. It expresses, psychologically, the joy of life and the life urge which overcome and eliminate everything dark and inhibiting. Where spring-like joy and expectation reign, spirit can embrace nature and nature, spirit.

—C. G. Jung.

The first time we open a hive there comes over us an emotion akin to that we might feel at profaning some unknown object, charged perhaps with dreadful surprise.

—Maurice Maeterlinck.

Introduction

From when I was in my early twenties, I worked for ten years as a beekeeper, helping my stepfather look after many hives before his death in 1983. I loved this work and it taught me much about the bees, the miracle of honey, and the way the hive functioned as an orderly busy community. A beekeeper can learn much about these things. However, the life and ways of bees, in spite of scientists having recently (October, 2006) published the honey bee genome (or gene map), still remain an ineffable mystery—that is, it can never be fully known to our consciousness.

****

The symbolism of the bee has carried numinosity for many centuries. It is vast, linking heaven and earth, god and man, the plant and animal kingdoms, sky gods and earth goddesses—including the supreme sun gods and the Great Mother Goddess.

Honey, and the nectar it is made from, is almost universally regarded as the food of the gods. Ever since humans first discovered the sweetness of honey and began to harvest the honeycomb, the bee has been venerated as a natural wonder.[1] Bees are social, industrious and beautiful, but also terrifying in that they have a powerful sting and venom and can act en masse. They can become very aggressive if their precious honey stores are threatened.

Honey, the miraculous sweet golden substance or healing elixir that is produced by the bees, can be seen as one form of the alchemical *lapis* (the living, flowing stone), while the bee's sting has been likened to that of the arrows of Eros/Cupid, the god of love. Symbolically, the bees' honey making, and their regulation of the hive (that is, of the whole colony) and dedication to their queen have much in common with the *chrysopoeia* (the gold-making) of alchemy and with the entire alchemical

---

1 The honeycomb is the wax hexagonal cells that contain the 'cured' or ripe honey.

opus, and thus with the psychological individuation process described by C. G. Jung. The symbolism of the bee, honey and the hive intertwine but at the core is the imagery of the circumambulation of the centre. For bees, this is a dynamic in the service of the queen bee and thus of *zoë,* the indestructibility of life. Psychologically it is linked with the mandala or archetype of inner order.[2]

The title of this book includes the words 'circumambulating the centre'. This focus on movement around a centre is not only fundamental to the human religious instinct, but also is quintessential to Jung's rich allusions to alchemical imagery, as he explores the process of psychological individuation in which the centre is named the Self or inner god-image.

My work of amplification will involve an approach to the alchemical union of opposites through symbolism related to the life and nature of the bee. Some particular opposites associated with the bee, honey and the hive include those of love and war, sweetness and bitterness, the individual and multiplicity, light and dark, order and chaos, industry and laziness, earth and heaven/the sky, sun and moon, spirit and matter, fertility and sterility, regeneration and death.[3]

****

I will first describe something of the biology of bees and consider them from the perspective of the natural sciences. Throughout I will honour, through literary work, the Muses so closely linked in mythology with bees. For example, Hesiod wrote: 'Happy is he whom the Muses love: sweet flows speech from his mouth',[4] and the Roman writer and expert on agriculture Columella wrote:

> *It is with good reason that the bees have been called the "Birds of the Muses", for if they are flying away, they can quickly be recalled by the*

---

[2] This comment is inspired by Karl Kerényi's subtitle for his book, *Dionysos: Archetypal Image of Indestructible Life.*

[3] There are many folktales, fairy tales, myths and legends in which the bee is central. The world religions often highly value the bee, the honey and the hive. Later in this work I discuss a selection of representative tales and motifs.

[4] Hesiod, *Theogony,* ll. 96f.

clapping of hands and tinkling of cymbals. Nature has also given the bees the undisturbed flower-covered hills for their habitation, just as men say that those divinities [the Muses] live on Helicon and Olympus.[5]

I will discuss the different castes and roles of bees within the hive and during their life cycles, as well as their pollination, nectar gathering and honey making as part of a chemical/alchemical process of transformation. I will also consider the stinging, swarming bee and the current crisis of Colony Collapse Disorder affecting bees in many countries.[6] In addition I will explore bees and honey symbolically in myths and legends, alchemical texts and various religions. Alchemical and psychological analogies will be presented, as I consider the individuation process through amplification of texts and images, in particular Niklaus von Flüe's vision of the 'Transformed Berserker'.

The latter is a vision that strongly links the golden bee and honey symbolism with love or deep eros that, in this case, shows that a Christian saintly hermit living in solitude can—through deep relatedness—accept and thus help transform the opposites of a wild collective situation and a warlike, but sacred, pagan animal side—that of a Wotanic bear sparkling with gold. Niklaus von Flüe (Brother Klaus) accepts a vision of one version of hell—another kind of 'Colony Collapse Disorder'—and is then able to emanate a healing political influence. His individual daily devotion or focus on a sacred centre is concentrated in such a way that war and disunity are not constellated in the collective. My work on bees was partly inspired by the deep eros that Marie-Louise von Franz evokes, as detailed later, when amplifying the symbolism of honey and love in von Flüe's vision:

Then Klaus realised that the wayfarer [i.e. pilgrim] bore him such love that he was quite stricken and had to confess that he was not deserving of this great love. Then he saw that this love was in the wayfarer. And he saw that his spirit, his face, his eyes, his whole body was full of this elevated love (Minne), like a vessel that is filled to the brim with honey. Then he could no longer see the wayfarer, but he was so fulfilled that he no longer

---

5 Columella. *De Re Rustica*, 111.16, 7.

6 See below, pp. 36f.

desired anything from him. It seemed to him that the wayfarer had revealed to him 'everything that was in heaven and on the earth'.[7]

Midway in my training to become a Jungian analyst, I had a dream in which I was travelling on the train from Lucerne, making the last stage of my journeying from Zurich to St Niklausen, near the place in Switzerland where Brother Klaus lived. This was geographically where much of my analytic training was centred. The dream told me my training was a pilgrimage. This image has encouraged me to maintain a pilgrim's stance, to keep my feet on the ground, but also to be mindful that there is a sense of the sacred and of a 'centre' in daily work and life. The bee's life is a circumambulation around such a centre—the hive protecting the queen—and has long inspired human beings of many cultures. In psychological terms, this describes the community of the Self as it manifests in the collective.

Biographical

In the 1970s, my stepfather Claude retired from the busy life of running a small hotel and spent time with an elderly relative named Walter—a wise and deeply reflective man who was approaching his death. Walter initiated Claude into bee lore and Claude took on the tending of several hives. I learnt to care for the bees by working with my stepfather and then, as he approached his own death, I was left to look after 150 beehives. This meant attending the hives in the forest, on orchards or farms. In summer, after the bees had filled the combs with honey, there were long hours of work in the 'honey house' at night. Tasks included uncapping the bees' wax seal on the cured honey with a hot knife (the uncapper), and melting down the wax before it could be sent to be rolled into sheets of foundation wax to be imprinted with the miraculous hexagon shape that assists the bees by providing a template for rebuilding their honey-storage vats or brood chambers.[8]

---

[7] Von Franz, 'The Transformed Berserker', in *Archetypal Dimensions of the Psyche,* p. 40.

[8] This six-sided shape is fundamental to the construction of any wax honeycomb the bees make. In the hive, as well as in organic nature, the hexagon is the key building block. What is called the 'foundation comb' adds to the 'archetypal' image of the hexagon, the

The moveable frames holding the dripping honeycomb were loaded into a huge centrifugal extractor and the golden liquid was spun out. Then, with a turn of the tap on the vat, it would flow into the 40-gallon containers. Impurities would usually be strained (sieved) out at this stage but some people would insist on having the unstrained honey with, for example, pieces of pollen or bee legs, as they believed the unstrained honey offered added nutrients.

We had to work with dedication and focus to extract the honey from the moveable frames of honeycomb, in order to return them to the hives without delay. 'Supers' (upper boxes), taken from the hives one day, would be returned to the bees in a day or two with foundation comb acting like hexagonal scaffolding ready for the bees to build up the honeycomb before refilling the extended hexagonal 'vats' or cells with honey. Grilles called queen excluders were used so that the queen, the longest and largest bee in the colony, remained in the nucleus of the hive.[9] That way, the brood box, housing the queen and her retinue, would remain undisturbed in the apiary.

Being a beekeeper, an apiarist, includes shifting the bees to be near new honey sources. The moving would take place during the dark of night when all the bees had returned to the hive and we could close up the entrances. (To attempt to move the bees under the light of a full moon, for example, would mean lost bees and the potential of many stings.) We transported bee boxes to various forests, following the honey flow (as it is called) or the seasonal blossoming of various eucalyptus trees, the native trees of Australia. Sometimes the hives were placed in

---

basic structure and shape of all honeycombs, a pattern which is innate for the bee. The basic structure of the honeycomb is the hexagon. So the symbolism of *six* (6) is important to consider. In *The Practice of Psychotherapy*, Jung comments that, according to old tradition, the number 6 means creation and evolution, since it is a *coniunctio* of 2 and 3 (even and odd = female and male). (CW 16, par. 679) Philo Judeas therefore calls the *senarious* (6) the 'number most suited to *generation*'. It represents the perfect division of the circle by its own radius. (CW refers throughout to C. G. Jung, *The Collected Works)*

9 This practice is one of several that may have, over many years, contributed to the current crisis in bee populations in Europe and North America. The publication of the honey bee genome and the announcement of Colony Collapse Disorder in the world media were synchronous events in late October 2006.

orchards to help pollinate the blossoms on the fruit trees.

Bees are nature's great pollinators, so they play a crucial role in the biological web of life. They also have a central role as culture bringers and psychopomps in the creation myths of many lands and in folk stories concerned with regeneration and *zoë.* Scientists are now honouring this pivotal role of the bee in nature in a different way, especially since 2006 when the honey bee genome was published, offering new insights into how vital the bee's instincts and the life of the hive as a whole could be in serving and educating human beings and in healing some of the ills of humankind.

Bees *(Apis mellifera)* and flowers are closely linked in nature. Flowers are essential as sources of nectar and pollen for the bees in their honey making. It is the often-golden centres of the flowers that attract the bees. In return, bees are essential for fertilising plants—pollinating flowers, cross-pollinating fruit trees and doing much to ensure the setting of the fruit.[10] This world of the flowering and fruiting (a particular kind of regeneration) is also central in alchemical symbolism as expressed by Cleopatra to Komarios in the first century A.D.:

> *Observe the nature of plants and from whence they come. Some come down from the mountains and grow up out of the earth, others rise up from caves and from the plains. But observe how one approaches them. One must gather them at the right moment, on the appropriate days. Pick them from the islands in the sea and from the upper plains. And observe how the air serves them, how the wheat embraces them protectively so that they are not damaged or destroyed. Observe the divine water which nourishes them, and how the air rules over them after they have incorporated themselves into one substance.*[11]

---

[10] Each day, other than in the cold of northern winters, sees the worker bees flying purposefully in search of nectar sources. Nectar, traditionally known as the 'food of the gods', is the raw material for the ultimate food of the gods, honey: a golden, sustaining, purifying and healing thick liquid. Honey bees and fruit culture must go together. 'Without this total contribution by bees, we would certainly live in a very different, less productive and less interesting world.' (from Langstroth, *On The Hive and the Honeybee).*

[11] *Komarios to Cleopatra,* in Marie-Louise von Franz, *On Dreams and Death,* p. 27. Von Franz also mentions here that, preceding this quoted passage in the *Komarios* text (one of the oldest texts of Graeco-Egyptian alchemy), there is an account of the entire

Here Cleopatra compares the central alchemical process with the growth of a plant.[12] The growing plant symbolises the mysterious, numinous, central thing that the alchemists concentrated all their attention on. This is like the worker bees whose life energy revolves around their queen, the honey-making process and the welfare of the hive. A growing plant uses the substance of the four elements—fire (the heat of the sun), water, earth and air. It combines them into the one thing ('the monad': the new living thing, namely the plant)—just as alchemy produces one substance drawn from all the surrounding substances.[13]

****

I enjoyed my work as a beekeeper and the solace and contrast it offered from days spent teaching, mainly adolescent boys, English at a technical high school. To work with the bees required an observance of a certain other rhythm of life. Tending the bees took me into nature and meant that I saw which flowers were budding or blooming at any particular time. I learnt about trees, seasonal changes and how crucial the early spring pollen sources are for nourishing the bees and for natural cycles of flowering and fruiting in general.

Commonly despised weeds, like the archetypal 'stone the builders rejected',[14] that farmers and graziers as well as government Departments of Agriculture wanted to eliminate, were in fact key elements for the building-up of the beehives' strength in spring. Apiarists fought to protect these weeds from being eliminated through overemphasis on crop monoculture or by being sprayed with weedicides. There was clover, the source of a thin blue honey. Another more controversial example is the

---

cyclical process through which the 'philosopher's stone' or gold is to be produced—a cycle in the form of a mandala of two times four colours and processes. Indeed, in mythology the blooming of many flowers is a central symbol that marks the *hierosgamos*: of Zeus and Hera, for example in the *Illiad.* Book XIV.

[12] To the Egyptians in Pharaonic times, the germination and growth of a plant, like wheat or flowers, symbolised the resurrection of the mummified body.

[13] Marie-Louise von Franz, *Creation Myths*, p. 227.

[14] The ordinary and rejected stone is like the alchemical *lapis* found everywhere (that is, it is a central part of ordinary existence). This 'orphan' turns out to be an important building block, as a foundation stone, or even a sacred stone.

beautiful purple-flowering plant that is deemed to be a 'noxious weed' in my home state of Victoria but was allowed to flourish in New South Wales, the state directly to the north. Farmers and graziers call it Paterson's Curse (because it bloats the cattle when they eat it greedily), but beekeepers call it Salvation Jane. Cape Weed, a big yellow daisy with a black centre, was another of these spring 'weeds' that provided the bees with vital sources of nectar and pollen to help them regain strength and numbers after the winter (and, by doing so, regain 'productivity', that modern 'buzz word'—a term that originated in the buzzing of busy bees).

To work with bees is to be actively aware of and immersed in the diurnal and seasonal rhythms, and to pay close attention to nature. The beekeeper, however, does not ultimately control, own or domesticate the bees. In spite of human attempts to breed hardworking, quiet and stingless varieties, essentially the bees retain the wildness of nature. The history of beekeeping 'has not been a story of domestication, but rather one of humans learning how to accommodate the needs and preferences of the bees themselves'—of nature.[15]

---

[15] J. L. Gould and C. G. Gould, *The Honey Bee,* p. 17.

1
The Bee in Nature and the Natural Sciences

Marie-Louise von Franz comments that the lives of all animals, 'even in the very early level of evolution, are shaped by what the behaviourists would call patterns, a certain mode of fighting, of courting, of looking after the young, of mating and so on which differs for each species'.[16]

The following section will consider some of the patterns and instincts of bees in order to provide a basis for understanding and amplifying the symbolism of the bee, the hive and honey.

There are three castes of bees in the hive: workers, drones and the queen (Figure 1). Bees are most industrious creatures. Each day (except during the coldest months of winter in the north), in response to sunlight, the workers start flying at sunrise from hive to flower or nectar source and continue to work all day.

Figure 1. *(left to right)* Worker, drone, queen bee.

---

[16] Von Franz, *Shadow and Evil in Fairy Tales,* p. 179.

The Queen Bee and the Hive

Bees are associated with hardworking, communal activity in the service of the future of the hive. This activity focuses on the queen, the only egg-laying female. The workers are underdeveloped, asexual females and the drones are the males.

Simon Buxton describes the queen and captures some of her symbolic and numinous significance poetically:

> *The queen [bee] is truly a queen in the ancient tradition. She is longer than her daughters [the workers], more slender than her sturdy sons [the drones], and mated with her own brother. In all her long lifetime [more than six to eight times that of the longest-lived worker or drone] she sees no other bees except her own sons and daughters. In her solitary person she carries the destiny of all, and she is a goddess whose life is devoted to selfless service within the dim light of the golden city.*[17]

The symbolic significance of the queen is enhanced by the facts that she lays some eggs by a process of parthenogenesis before she has been fertilised by a drone; that these eggs result in drones; that the queen mates only during her few flights away from the hive, with carefully selected drones; and that during a single encounter, the drone provides enough semen for the queen to lay tens of thousands of eggs. From then on, the queen's life is a process of never-ending egg laying after which the brood or larvae that develop can become either workers or queen depending on what food the nurse bees provide.

Modern scientists have noted that one of the hallmarks of the communal life of *Apis mellifera* is behavioural differentiation among individual worker bees, which results in a division of labour[18] in the hive or colony that serves the life force of the whole and enables the bees, as Virgil says in the verse given here later, to 'all' (that is, the multiplicity of bees) act with 'united force'.[19]

Almost all the bees in a colony or hive are female: a single queen bee

---

17 Buxton, *The Shamanic Way of the Bee,* p. 25.

18 Gene Robinson, 'Hormonal and Genetic Control of Honeybee Division of Labour', in L. J. Goodman and R. C. Fisher, *The Behaviour and Physiology of Bees*, p. 14.

19 Virgil, *Aeneid,* I. 433. See below, pp. 20f.

lays up to 2,000 eggs a day, and the tens of thousands of workers in the hive are female. Worker bees, though, have poorly developed ovaries and cannot mate. In cooler climates a few hundred (or more in a large hive) male bees or drones are reared for part of the year. The large furry drones, whose name has long been synonymous with laziness, exist only to mate with the queen.[20] The drones mostly stay in the hive and are bigger and much lazier than the worker bees. Drones are needed in summer time, as they are the inseminating force in the hive. Their raison d'etre is the possibility of mating with the queen. In winter, because they are unproductive and only would eat the precious honey store and create more for the workers to do, when sacrifices have to be made for the good of the community, the drones are stung to death by the queen and the workers who then throw them out of the hive.

The worker bees build the honeycomb based on a natural template of hexagons and, in their roles as nurse bees, they rear the young bees. The workers feed the queen a special gel-like substance called royal jelly; they also feed the drones, guard the hive against robber bees and other intrusions, and collect food (pollen, nectar and water). It is the worker bees who give the colony its reputation for industry, order and dedication to daily rhythms that enhance the future life of the hive and its epicentre, the queen.

Worker bees (the asexual females) work very hard to bring in the raw materials for the whole hive and for the honey-making process. They guard the entrance to protect the egg-laying queen and the whole hive with its nursery and brood chambers, honey stores, walls of waxen honeycomb in which gaps opening to the outside elements are sealed with propolis (bee-produced gum). The task differentiation, and what the scientists also term 'altruism,' in a beehive is now being researched to unlock mysteries that might have application to human social behaviour.

In a hive some worker bees are nurses who attend the brood, others are foragers and guards ('sentinels'), but all have the queen as the central focus of their daily work. The whole energy of the bees is directed to

---

20 The theme of the Grimms' fairy tale 'Lazy Harry' is laziness. *(Grimm Brothers, The Complete Grimm's Fairy Tales,* pp. 678f.) The bees, the honey and the hive all feature in this tale. See also von Franz, *The Feminine in Fairy Tales,* pp. 62f.

serving this centre, a ruling principle around which the entire life of the hive is focused. In the Grimm's fairy tale 'The Queen Bee',[21] the queen is drawn to the pure honey, which is only found on the true princess's lips, thus differentiating the real princess and future queen from the false ones. This suggests that the queen of the hive points the way to or symbolises the life-preserving, inspiring and nourishing eternal value of what, in depth psychological terms, is called the Self.

In the symbolism of divine inspiration we often see the image of a bee settling on the honeyed lips of the gifted mortal. In this way the honey bee is connected with the descent of the gifts of the spirit to humankind and then, in return, with the potential offering-up (the ascent) by humankind of creative work with eternal value. In answering the question, 'What is a poet?' Elizabeth Henry wrote:

> *He is a man of religious experience whose creative gift enables him to communicate spiritual truths to men. His poetry can bring deliverance from spiritual death, bringing his hearers to a new knowledge of their divine Creator, who gave him this special power. In this way souls that have been disordered can be healed, and the human relationship with God may be restored when it has been impaired. . . . This is the fruit and indeed the purpose of music and poetry, direct gifts from God to mankind.*[22]

The poet Virgil lived from 70 to 19 B.C. The following extract from his *Aeneid* shows that the remarkable division of tasks among worker bees in the hive has long been appreciated:

> *Such is their toil and such their busy pains,*
> *As exercise the bees in flowery plains.*
> *When winter past and summer scarce begun,*
> *Invites them forth to labour in the sun.*
> *Some lead their youth abroad, while some condense*
> *Their liquid store, and some in cells dispense:*
> *Some at the gate stand ready to receive*
> *The golden burden, and their friends relieve;*
> *All, with united force, combine to drive*

---

21 *The Complete Grimm's Fairy Tales,* pp. 317f.

22 Henry, *Orpheus with his Lute: Poetry and the renewal of life,* p. 62.

The lazy drones from the laborious hives,
With envy stung they view each other's deeds;
The fragrant work with diligence proceeds.[23]

Worker bees, which make up the bulk of bees in a hive, have a four- to seven-week adult life that typically involves a progression from nest activities, during the first two or three weeks, to foraging. There are four behavioural phases or 'age castes', each of which is devoted to the performance of several different tasks. Bees are sensitive to environmental influences and can vary these tasks to match the needs of the colony—the 'unity in multiplicity' that is the hive. This is a kind of self-regulation of the whole, and it is self-regulation that Jung speaks of in connection with the healing of the psyche through the Self, also reflected in the kind of healing mentioned above by Elizabeth Henry.

The bees' order, efficiency and hardworking dedication to the central tasks of keeping the hive flourishing have long been awe inspiring to humans. In particular, the social organisation and instinct of bees has been cited over the centuries as a fine example of an altruistic highly ordered and self-regulating community of miraculous creatures devoted to serving their queen. In religious and political contexts (spheres which were identical or overlapped considerably in past eras), the bee has long been venerated as a role model for industry, community, order, chastity, and devotion to duty. For example, in 1578 Guillaume du Bartas wrote: 'Where's the state beneath the firmament / that doth excel the bees for government?'[24] and Samuel Purchas in 1657 commented that the bees demonstrated the 'invincible' power of the multitude, and that they would 'conserve community until their last'. He added:

Bees are political creatures and destinate all their actions to one common end; they have one common habitation, one common work, all work for all, and one common care and love towards all their young, and that under one commander [the queen bee].[25]

---

23 Virgil, *Aeneid,* I. 430–436.

24 Du Bartas, *Divine Weeks and Works.*

25 Purchas, *A Theatre of Politicall Flying-Insects,* p. 16.

In 2006, we have the following:

> *When compared to other insects, the honey bee genome contains fewer genes involved in innate immunity, detoxification enzymes, and gustatory (taste) receptors, while not surprisingly, it contains more genes for olfactory receptors and novel genes for nectar and pollen utilization. Interestingly, the honey bee genome shows greater similarities to vertebrates than insects for genes involved in circadian rhythm, as well as biological processes involved in turning genes on or off.*[26]

It is also important to remember that bees have an aggressive side when they sense a threat to their queen, hive or honey. They have a potent, painful sting, and when they use it they sacrifice their lives.[27] In 1767 beekeeper John Mills stated: 'The sting is very necessary for the working bee, both as an offensive and as a defensive weapon; for their honey and wax excite the envy of many greedy and lazy insects'.[28]

A bee's sting has a hook-shaped barb, like half an arrowhead. To remove an embedded sting cleanly, you need to dislodge this natural arrowhead. The injected poison produces immense heat in most people's bodies and, often, significant swelling. Some humans have a strong allergic reaction to the sting, and in some cases the venom can be fatal.

Bee venom can also be a source of healing. Controlled bee-sting therapy is used in traditional and alternative medicine, and other substances produced by the bees have for a long time been used medicinally. Penotus, in *Theatrum chemicum I*, wrote: 'The elixir of honey preserves and cleanses the human body from all imperfections, both within and without',[29] and in homeopathy, for example, where in the right measure 'like cures like', the *Apis* remedy is made from the whole bee, including the sting. One of the main uses of this remedy is to treat the stings of other insects, and for *urticaria* (commonly called 'hives') and other instances of burning or itchy skin that may be swollen or sensitive.

---

26 On the publication of the honey bee genome, in *Nature: International Weekly Journal of Science,* no. 443 (26 October 2006).

27 For example, a June 2007 headline in Melbourne's *The Age* was 'Killer Bees'.

28 John Mills, 'An Essay on the Management of Bees, Wherein is Shewn the Method of Rearing those Useful Insects', pp. 8f.

29 Quoted in *Mysterium Coniunctionis*, CW 14, par. 683, n. 81.

Propolis, the golden–brown gum or glue that bees use to seal little gaps in the outer structure of the hive, is a natural antibiotic and is helpful in healing wounds, as is the direct application of honey itself. Sealing using propolis is the natural way the bees protect the contents of their hive from the elements. This evokes the imagery of the well-sealed vessel or retort in alchemy, or the temenos of a Jungian analysis—the container of a work of transformation and regeneration. Edward F. Edinger talks of the importance of such a substance in alchemy. The gum or 'glue' is adhesive and holds things together; thus, in alchemy, glue, gum and resin are synonyms for the transforming substance called by some the 'glue of the world' *(glutinum mundi).*[30]

Royal jelly (a creamy gel that is secreted by the worker bees to feed young queen bees or to transform the brood or larvae into the stuff of queens) and pollen are also used in natural healing.

****

Like the ant, the bee is what scientists call a 'eusocial' insect and, as such, the bee is a symbol of self-discipline, social organisation and ceaseless toil. But it is also productive and creative: it collects nectar from sweet-scented flowers and converts this liquid into the paradoxically viscous yet flowing honey—the golden and enduring miracle food and immortal nourishment of the gods. The beehive is a symbol of buzzing, well-organised productive efficiency in which every individual performs an allocated task in 'selfless' service of the queen. The hive is thus widely regarded as a symbol of a well-run, prosperous state, and the bee shares in the symbolism of the culture-hero who brings civilisation and order through wisdom and example rather than by force of arms.

As in the Grimm's fairy tales 'The Valiant Little Tailor' and 'The Queen Bee', the ants and the bees, insects that live in highly organised communities, are focused on the goal. And, in the case of bees, this goal is service to the queen who lays the eggs and keeps the hive alive. At any given time, every individual bee in the hive has its particular task, and there is the sense of what could be deemed a regenerating 'conscious'

---

30 Edinger, *Anatomy of the Psyche,* p. 223.

teleology. The bees respond to sunlight and moonlight, and use colour and odour differentiation. Their innate sense of orientation is legendary, and they have the ability to find sources of nectar and pollen and to communicate in a form of language to the other bees in the hive where these sources are, how potent they are, and in what direction the worker bees will have to fly to gather the nectar and the pollen. There is also an indication of the intensity of the honey flow or nectar indicated by the vigour or otherwise of the tail wagging or circular dance of the bees.

Recent scientific findings published since 2006 as a result of the work in and following mapping of the honey bee genome have reinforced the long-held view that the bee is a truly remarkable insect and have revealed that bees are closer to the vertebrates than previously thought, and closer than we had thought to having consciousness. There has been much debate about this. In his article 'Synchronicity: an acausal connecting principle', Jung explored this issue as early as 1952, saying that given Karl von Frisch's research into the bees' communication through dance language proved the existence of transcerebral thought and perception, 'we are still faced with the fact that the ganglionic system [in bees] apparently achieves the same result as our cerebral cortex. Nor is there any proof that bees are unconscious'.[31]

Nectar, Honey and Tapas

As in the gold making of alchemy and in the psychological individuation process we start with unrefined *prima materia*—the raw affects and 'excrement'—the *ultima materia,* the honey, if described unromantically, begins as regurgitated nectar and aphid droppings. These raw materials are then transformed by enzymes (chiefly invertase) in the bees' glands into an 'invert' solution, which is then evaporated in the wax cells into a concentrated syrup through a wing-fanning action. Contemporary writer Bee Wilson has described the real 'secret' of honey reductively as chemical- or enzyme-action plus hard work. The deeper and more mysterious secret, however, when honey making is considered symbolically as an elixir making or *lapis* making alchemical process, is that of the

---

[31] *The Structure and the Dynamics of the Psyche,* CW 8, pars. 956f.

'Golden Flower'—of the symbolic sacred, life-sustaining and regulating centre and its nourishing and healing properties.[32]

Nectar, the main raw material of honey, is a more or less watery solution of sugars that vary depending on the flower source but always include sucrose. Worker bees, from when they are three weeks old, know how to travel up to three kilometres (in some rare cases, double that) to reach a good source of nourishment (nectar or pollen). They work remarkably hard and with great dedication. They may visit as many as a thousand flowers on each foraging trip, sucking nectar out of the centre of each flower and carrying it in their honey-stomachs, before returning to the hive to deposit the load onto the proboscis of house bees who then disgorge (vomit) the sweet liquid into the hexagonal waxen cells.

In the process of converting nectar into honey, worker bees keep the inside of the hive at a temperature that helps to evaporate excess water and to concentrate the sugar content of the liquid which is to be transformed into honey.[33] The bees' hypopharyngeal glands secrete the enzyme invertase, which inverts most of the sucrose in nectar into fructose and glucose. Fructose is more soluble in water than glucose or sucrose, and as a result of the relative solubilities of the sugars in a solution containing all three (at temperatures in a honey bee colony) an extra total amount of sugar can be held in solution and more water can be evaporated. The maintaining or regulation of the brood heat in the hive is like ensuring tapas, a ritual heat about which Jung writes in his alchemical studies.[34] The temperature of transformation and incubation in the hive well sealed with propolis, or bee gum, is kept just right (at approximately human body temperature), just as the heat of fire or level of eros needs constant attention and regulation in the alchemical vessel or the analytical temenos. As the alchemist Komarios says: 'In putrefaction and

---

32 See von Franz, *On Dreams and Death*, p. 33; also *Mysterium Coniunctionis*, par. 714, in which Jung says Mercurius is the *prima materia*.

33 Eva Crane, *The World History of Beekeeping and Honey Hunting*, p. 21.

34 See *Mysterium Coniunctionis*, CW 14, par. 262, n. 487. 'Tapas' means self-incubation (brooding in the *dhyana* state) and is described as a self-heating or incubation where one ponders over or broods over the unconscious contents and digests them. See also von Franz's comments on tapas in *Creation Myths*, pp. 130f., 143.

warmth, there is created a remedy which permeates every body without hindrance.'[35]

The bees flap their wings to keep the inside of the hive at a high-enough temperature. Within this temperature-regulated vessel, the liquid cures or ripens to the appropriate consistency within the wax 'vats'. The bees then cap or seal these small hexagonal cells with extra wax. This helps keep the honey fresh during the winter period when, if it is not harvested by a beekeeper, the bees will need to feed from it.

The chemical process in honey-making ultimately produces a natural golden substance that does not deteriorate, a sticky substance that flows and washes off easily in water, that can crystallise but then regain its fluidity with a little heat—in other words, a pure or unalloyed health-giving elixir with an eternal value. Doctors and healers, as well as their patients, have marvelled throughout the ages at the efficacy of honey, not to mention its most widely acclaimed quality—its delicious sweetness. The high total sugar concentration in honey is beneficial in that most yeasts cannot ferment in it. With the natural addition of one other constituent (glucose oxidase), the honey develops antimicrobial properties, and it can be stored safe from spoilage. This accords with what the hive of bees needs, as the concentrated finished honey (or, as it is called, the 'cured' honey) also occupies less storage space in the nest, allowing more room for egg laying and brood rearing.[36]

Worker Bees

Spring and Kundalini Rising

In March 2006 at St Niklausen, the catkins hung from the willow trees and the other early blossoming trees were filled with the buzzing bees. This buzzing was so loud; I had never heard anything quite like this before, even though I worked with so many beehives placed together in the apiary. Maybe this is because I live in Australia where the bees can work

---

35 Andreas Schweizer, 'The Teaching of Komarios and Cleopatra: Reflections on a Greek alchemical text dedicated to Komarios and Cleopatra', (lecture, ISAP, Zurich). See also Schweizer 'Observe Nature and You Will Find the Stone: Reflections on the Alchemical Treatise "Komarios to Cleopatra" ', pp. 81f.

36 Crane, *The World History,* p. 22.

in winter and, even if the spring blossom comes after a relatively long cold winter, then the need for pollen gathering to build up the strength of the hive is not so urgent. The intensity and focus of the pollen-gathering work of these northern bees after a particularly heavy, snowy Swiss winter made a great impression on me and several others who had witnessed the trees buzzing with bees or bees flying with full pollen baskets, showing as ovals of gold on their little legs. A synchronicity was that at this time we also heard in lectures from Charles Zeltzer about the rising Kundalini symbolised by a serpent and the buzzing noise that emanates like the sound of bees when the sleeping Kundalini fire is activated and begins to rise from the base chakra. In traditional Tantric yoga, it is said that the sleeping Kundalini emits 'a sweet murmur like the indistinct hum of swarms of love-mad bees' and that 'she makes this sound when awakened'.[37] This Kundalini rising is like the buzzing of bees in spring—a crescendo of energy and urgent potential, moving up from the earth to the heavens.

Summer: Navigation and Dance Language

The oft-used expression 'busy as a bee' celebrates the unfailing energy of honey bees and, as we have seen, the worker bees serve the queen, the feminine life-giving and life-preserving force in the hive. It is as if all their eros or life energy dynamically flows into their focus on this task—the world of the Great Mother meets that of the sun god. Spring time, the alchemists tell us, is the best (and natural) time to begin the alchemical *opus*. It is a work of inspiration by solar consciousness and creativity accompanied by natural flows and meanders, and a fertilising of this inspiration by the wisdom of the unconscious that results in a more lunar consciousness. It is interesting to note though at this stage, and I will explore the alchemical aspect later, that the beehive was not only considered to be an attribute of the Great Goddess as Demeter-Ceres-Spes, but that the bee was also associated with the moon. The priestesses of the moon, like those of the Mother goddesses, were called 'bees' or *melissae* and it was believed that all honey came from the

---

37 Gerhard Adler, *The Living Symbol*, p. 10, n. 75; also Arthur Avalon, *The Serpent Power*, p. 29.

moon, the hive whose bees were the stars.[38]

The worker bees' activities include scouting for nectar and pollen sources and returning to the hive to transmit that information to the other bees inside the hive by what has been called dance language. Nobel laureate Karl von Frisch devoted his life to understanding the bees and the way individual bees communicate with each other for the good of the whole hive. On returning from their flights outside the hive, the worker bees dance and waggle, moving in a particular direction and with a specific intensity depending on what it is that they wish to communicate.

When a good food source is discovered by 'scout' workers, they return to the hive. Shortly after their return, many foragers leave the hive and fly directly to the food. A remarkable thing is that the foragers actually do not follow the scout bees back—the scout bees may remain in the hive for hours after they have communicated the information of their dance. This indicates that the scout bees have communicated the necessary information to the forager bees within the hive so that they can find the food sources independently. This information concerns the odour of the source and its direction and distance from the hive. When the food source is within 50–75 metres of the hive, the scouts dance a circular dance, now known as the 'round dance', on the vertical surface of the honeycomb. However, when the food source is more than 75 metres from the hive, the scouts dance the 'waggle dance' (a figure of eight: the shape of the lemniscate or infinity symbol). The waggle dance includes a straight run, the direction of which conveys information about the direction of the food source. The speed at which it is repeated tells the bees how far away the food source is.

Figure 2 pictures bee researcher Karl von Frisch, who discovered how the dance language of scouts communicates to the other worker bees inside the hive about the current sources of nectar for the main task of honey making, and the circle dance and the 'waggle' dance.[39]

Von Frisch discovered much of what is known today about honey bee communication (and was honoured for this with a Nobel Prize in 1973).

---

[38] Robert Briffault, *The Mothers*, vol. III, p. 179.

[39] Karl von Frisch. 'Decoding the Language of the Bee', p. 78. See also Eva Crane, *The World History,* pp. 52, 568.

Figure 2. Bee researcher and Nobel laureate Karl von Frisch.
Bottom diagrams *(left to right):* bees' circle dance and waggle dance.

One important question concerned how the bees calculate distance. Von Frisch initially thought they measured the distance by assessing the amount of energy it took to get to their destination. But further investigation revealed that bees measure distance by the motion of images received by their flicker-sensitive compound eyes as they fly.[40]

The bees have a remarkable innate direction-finding ability. This is related to the position of the sun (symbolically the central solar principle of consciousness) in the sky. The sun is like the bees' compass or lodestone; they orient their movements according to solar principles. In spring and summer, the worker bees focus strongly on gathering nectar and pollen from the flowers, and bringing these back to the teams of bees in the hive for processing, or for storage in the case of pollen. Figure 3 by von Frisch illustrates various elements in the bees' direct finding abilities, including their dance communication.

Flexible task differentiation is a key way each particular hive embodies a constant process of self-regulating in the varying of the daily work of individual bees according to the needs of the totality—the hive.

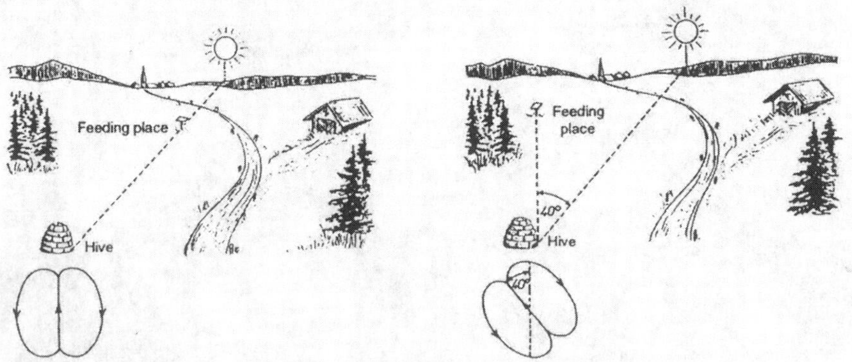

Figure 3. Indication of direction by tail-wagging dance. *Left,* the goal is in the direction of the sun; *right,* the goal is 40° to the left of the sun's position.

Dance figures, enlarged, are on the bottom left of the pictures.

40 Von Frisch, ibid.

At any time the presence of smoke is a signal for the bees to prepare for a fire and engorge themselves with honey from their stores and this, in turn, placates or dulls their desire to sting, which means the beekeeper using a smoker can open the hive and remove the heavy frames full with capped honey. They are taken to the centrifuge to extract the honey after the removal of the flat wax covering that seals in the honey. This process is called uncapping; the wax cappings are sliced off, usually with a heated knife. The melted wax is collected as another valuable resource provided by the bees.

Illustrating something of the imperative of self-regulation in the hive of bees, no second queen can exist there. The laying queen and her hive defend this principle utterly. Finger-joint-sized queen cells are longer than worker bee cells and house virgin queens or 'princesses'. If allowed to hatch out, they will fight so that only one queen lives. The workers will also sting the virgin queens to make sure there is only one living queen.

To requeen a hive, beekeepers also look for the queen in the open hive to kill her so that a new laying queen can be allowed to live and invigorate the hive.

Wintering: The Nigredo

The bees in colder climates cannot visit flowers for three months of the year. This relatively sunless and honeyless time is what beekeepers refer to as 'Wintering'. This is a time of suspended animation, of darkness and of suffering in holding the tension before the season changes, bringing with it a new consciousness, a sunrise or moment of spring-like regeneration. The dark time, like a person's depth psychological work with the shadow, akin to the alchemical *nigredo,* is not an easy time. Overtly, there is no work for the bees or the individual to do, but in fact it is a dangerous but potentially creative time of suspension between the opposites of work and inertia. There is a whirling in the beehive around a centre but this might seems like a void or still centre like the 'eye of a storm'. The opposites are not united: order and disorder, remembering and dismembering, love and hate, darkness and light, gold and black—

between life and death, inspiration and resignation.

****

Poet Sylvia Plath grew up with a difficult father who was a published expert on bumble bees. She herself became a beekeeper just before she wrote 'Wintering'. In this work she captures something of what happens for the beekeeper and for the bees inside the hive's darkness in winter.[41] In the cold months the bees maintain a constant temperature to keep alive a nucleus hive centred on the queen bee. Beekeepers often supplement the hives' honey supplies with sugar syrup, a less pure honey substitute that the bees feed on during winter. In the poem, animation is at first suspended and then regimented as the bees line up to sip sugar syrup. Joie de vivre is suspended—hanging on. Life is black but there are glimmers of golden light, and of consciousness, in the darkness, and there is the white of the sugar and the snow (like that of *solificatio* and the *albedo* in alchemy when light comes to the darkness). For the bees, there is a great resurgence of the life force in the springtime. The hive literally becomes 'a hive of activity'.

Swarming: Chaos and Interregnum—A Move Towards New Order

New bee colonies are formed by the bees swarming. The queen starts to lay eggs before winter ends. When spring arrives, brood rearing within the hive increases at a rapid rate. Then in early summer the population of bees in the colony can become so great (many thousands of bees) that drones (the bigger, lazy male bees) and several queens are likely to be reared instead of workers. This is the urge for movement towards forming new colonies. I will discuss this more when considering the queen bee and her mating or 'nuptial' flight in the *hierosgamos* section of chapter 5. When the new queens in their cells are nearly adult, a primary swarm (a cluster of about half the worker bees following the original queen) may leave the hive in search of a place to build their honeycomb and form a new hive with a new queen. The swarm usually will rest, for example, in a temporary cluster on small tree branches and scout bees

---

41 Plath, in *Ariel,* p. 63. Plath was experiencing her own deep 'dark night of the soul' before her early death when she wrote 'Wintering'.

will fly off looking for a suitable natural place (a cavity) big enough for a new nest. The volume of bees usually ranges from 20 to 100 litres.

A queen bee, larger and longer than others, is always meticulously attended by a retinue (the official beekeeping term) of worker bees. But as in fairy tales, those dreams of the collective, there comes a time of renewal when the old sovereign must die to allow for a new ruling principle or attitude to take its place in the collective. Thus, in depth psychological symbolism, the alchemical transformation or healing gives birth to the *filius philosophorum* or *lapis,* the Self.[42] The queen bee is a dynamic and strong symbol of the psychological Self. She lives in the darkness of the hive's interior but all the activity of the hive is regulated and adapts to serving her and the viability of the hive.

In the case of bees, the chaos or the swarm, for me, is linked with the symbolism of an interregnum or liminal period in the life of a particular hive or community of bees. Sometimes new queen cells develop and new queens or 'princesses' are allowed to hatch out. But, as I've already mentioned, in most instances the incumbent queen and the workers do everything they can to prevent this. If another queen does hatch out, it causes consternation in the hive. Self-regulation is then prompt and unequivocal—the other bees will act to stop this unstable situation progressing. Swarming is one way forward and this allows a new colony to form around a new queen, but usually the existing laying queen is preserved and order is restored within the original hive.

When approaching the bees, an attitude of respect, if not reverence, and an appreciation of the integrated functioning of the hive are needed, otherwise anarchy and belligerence, swarming and stinging, can break out. This swarming resembles a human psychosis, when an individual is overwhelmed or flooded by the contents of the unconscious. Insects in general are creatures that are so 'other' that they are often a component of unsettling visions or hallucinations in psychotic states or drug addiction. In addition, killer bees have been the subject of horror movies, as a

42 *Mysterium Coniunctionis*, CW 14, par. 14. Jung here writes of the 'virginal or maternal quality of the *prima materia*, which exists without a man and yet is the "matter of all things" '. He says that, above all, the *prima materia* is the mother of the *lapis* or *filius philosophorum*.

bee sting or several bee stings can be lethal to human beings. On the other hand, beekeepers and those who know and respect the bees can maintain a psychological calm in their presence. Bee Wilson writes:

> *The power to master fear, and even to love the thing they ought to fear. Some beekeepers claim they don't get stung as much as other people. . . . Stings or not, beekeepers can walk among the bees more easily than among men, they understand the power of stillness. In human life, self-effacement makes you fade out of view: but, in the apiary, self-effacement makes you strong.*
>
> *. . . While they are in the eye of the swarm, they seem curiously invulnerable . . . more than human because they seem to have mastered the secret of the hive, and therefore the secret of nature itself.*[43]

****

Swarming is the beehive's equivalent of a renewal as in the ancient Egyptian Sed festival in that it marks an interregnum period. Von Franz writes inspiringly about the old king in fairy tales who is sacrificed or has to go into exile.[44] C. A. Meier (commenting on the psychological parameters of examinations) said that the history of a particular term or word often discloses important information about its psychological background. He went on to say that the Latin word *examen* 'connotes a swarm of bees leaving the hive . . . because this exodus is equivalent to establishing a new status, which in turn represents the main object of every examination'.[45]

However, leaving the old status amounts to a separation and that has its dangers. According to Jung, it is an archetypal experience. It represents a critical phase. Each of us passes through several critical phases or

---

[43] Bee Wilson, *The Hive: The Story of the Honeybee and Us,* p. 233.

[44] Marie-Louise von Franz, *The Cat*, p. 24. In the Sed festival, the ruler was symbolically sacrificed and renewed in a psychological death and rebirth ritual, a variation of what in former times would have been an *actual* sacrifice or killing of the ruler for the renewal of 'the highest value' [that is, the king or queen] in the collective. See also D. Stephenson Bond, *The Archetype of Renewal: Psychological Reflections on the Aging, Death and Rebirth of the King.*

[45] Meier, 'The Psychological Parameters of the Examination Situation', in *The Well-tended Tree,* pp. 49f.

Figure 4. Some methods of taking a swarm, according to where it had settled. (E. Aphandry, 1911.) *Above:* (left) An accessible branch where a swarm had settled was shaken to dislodge the bees so that they dropped into a skep held below; (right) A low branch was cut on each side of a swarm clustered on it, and the swarm was carried on the branch to the new hive. *Below:* (left) A swarm on a vertical surface was brushed onto a container held close below; (right) A skep might be fixed above an inaccessible branch, and the bees driven up into it by a vigorous smoking.

So the swarm is such a turning point in the life of the beehive: it offers the potential for the new, but in the meantime some kind of chaos breaks out. As Andreas Schweizer pointed out to me while I was in the process of this writing,

> *Although the bees [symbolically], as any other insects, can indicate some sort of chaos or dissociation, there is always a secret order in this chaos. It seems that it depends a lot on the attitude of consciousness in which way this symbol develops, whether more towards chaos, or towards order.*[46]

This swarming energy leads me now to mention the most recent form of chaos affecting bees in commercial apiaries in Europe and North America, a syndrome known as Colony Collapse Disorder.

Colony Collapse Disorder

Turning and turning in the widening gyre
The falcon cannot hear the falconer,
Things fall apart: the centre cannot hold;
Mere anarchy is loosed upon the world,
The blood-dimmed tide is loosed and everywhere
The ceremony of innocence is drowned;
The best lack all conviction, while the worst
Are full of passionate intensity.[47]

Disorder and disorientation are strongly evoked by William Butler Yeats in the above words that start his great poem, 'The Second Coming'. In terms of the bees, we can speak of these states as the opposites of order and the innate orientation to the hive and its centre, the queen.

****

One third of the world's food is currently produced with the help of bees as pollinators. But since late 2006, headlines throughout the world have brought collective attention to the fact that hundreds of millions of bees in North America and Europe are dying or losing their capacity to find their way home to their hives. Many hives are almost empty of bees and there is honey unrobbed in these hives. This is honey that for some mys-

---

46 Personal communication.

47 From Yeats, 'The Second Coming'.

terious reason is not attracting bees. Normally, if there were a weak hive, robber bees from other hives would come in and take all the honey for their own nourishment and hive. This honey that is left is just not attracting the bees. What is wrong with the honey and what imbalances are causing the bees to lose their natural way?

These problems seem to indicate a kind of burnout: overwork of the bees and human irreverence for the natural needs and processes of these insects that are so vital to the welfare of the planet. The mystery remains but perhaps the applying of too much commercial ego consciousness could underlie this. This debilitating disorder came to global consciousness at the time of the publication of the honey bee genome in late 2006.

That the bees are not finding their way home and that honey, the miraculous food, seems to be contaminated are indicators that something unnatural has transpired. Einstein reputedly said that if the bees die, then without these pollinators the globe would be in crisis within four years. 'If the bees disappeared off the surface of the globe, then man would only have four more years of life left: no more pollination, no more plants, no more animals, no more man.'[48] And more recently Zac Browning of the American Beekeeping Federation reminded us that 'every third bite we consume in our diet is dependent on a honey bee to pollinate that food'.[49]

The energetic, orderly bees that have long provided humans with such models of natural structure and altruism are showing us that even this can break down into its opposite. But what does remain is the 'nucleus' hive around the queen as centre. This does not die. A hive is naturally self-regulating but ultimately, like the psychological Self and the wisdom of the collective unconscious, it remains a mystery that resists being harnessed to the ego's and the collective's purposes; for example, research scientists, pharmaceutical and pesticide manufacturers, and large-scale highly commercial apiaries or pollination services.

It remains to be seen how nature or science can transform or restore balance to this challenging situation. The sudden and mysterious death of

---

48 This now well-known quote is attributed to Albert Einstein.

49 Browning quoted in Christine Evans, 'Mysterious collapse of honeybee populations threatens national food supply'. (www.naturalnews.com)

the bees does, however, have an archetypal precedent in the sudden death of the ancient Greek god Aristaeus's bees. The myth of Aristaeus (retold, for example, by Virgil in his fourth *Georgic)* points to a sacred solution (involving the creation of a restorative mindfulness or new religious attitude), as does the Transformed Berserker vision of Niklaus von Flüe. I shall discuss both of these later.[50]

Beekeepers

> *These are the magicians who extract 'honey from the weed' . . . the weather-hardened farmers who have learned how to find happiness in modest circumstances.*[51]

Humans, in spite of all their efforts, have not really succeeded in domesticating bees. If a swarm escapes from a commercial apiary, for example, it will have the same chance of surviving in the wild as a feral swarm. The number of wild colonies living in trees still far exceeds the bee populations living in accommodations designed for them by humans. Indeed, the history of beekeeping has not been a story of domestication, but rather one of humans learning and paying attention to the needs and preferences of the bees themselves. We have much to learn from the natural life of the bees.[52] Karl von Frisch and many traditional beekeepers knew this, as do contemporary bee-research scientists like Gene Robinson and May Berenbaum and their colleagues in their work at the cutting edge of honey bee genomics, neurobiology and entomology.

Bee Wilson differentiates what she calls three archetypal kinds of beekeepers: the scientist or researcher, the showman or magician, and the hermit or sage. The magician type is a Mercurial figure performing tricks

---

[50] The headline news of Colony Collapse Disorder threw me and many others off centre, and for me engendered doubts about the ultimately self-regulatory capacity of the bees and thus of nature. I am grateful for this experience as it helped me avoid onesidedness, and returned me to a renewed sense of an inner centre. This was like 'new sun' stage (as Jung says, 'a new knowledge'). I could return then to my writing about the Transformed Berserker having a new knowledge from a lived experience of this particular 'problem of the opposites'.

[51] Wilson, *The Hive,* p. 233. The quote is from Shakespeare, *Henry V*, iv. i.11.

[52] Gould and Gould, *The Honey Bee,* p. 17.

like getting a swarm to settle beard-like on his chin.[53] A scientist is one who researches the secrets of bee lore. The sage is someone who leads a quiet and often contemplative life and, through bees, submits to nature.[54] As bee-man and novelist Tickner Edwardes's bee-master says:

> *The bees had their definite plan of life, perfected through countless ages and nothing you can do will ever turn them from it. . . . You can delay their work or you can even thwart it altogether, but no one has ever succeeded in changing a single principle in bee-life. And so the best bee-master is always the one who most exactly obeys the orders from the hive.*[55]

Many beekeepers can work or be close to bees without protective clothing. They are often unassuming people who somehow blend in with nature. There is a sense of mystery surrounding the traditional role or work of the beekeeper. Domestic bees in constructed hives rely on a human keeper to assist them fulfil their natural function. This role was a numinous one in past eras when, for example, if a beekeeper died the hives would be draped with black crepe and it would be the solemn duty of a living member of the family to 'tell' the bees their master had died and ask them to keep on working. The 'telling of the bees' was a necessary rite that had to be administered, otherwise the bees might not work for their new 'owner' or they too might soon die.

The role of beekeepers has increasingly become one that is out of balance (out of tao) and apiaries have come to be regarded as commercial production systems. This lack of eros between the bees and the beekeeper can engender the stresses of burnout and overwork in the commercial apiary, which now seem to be manifesting in symptoms like reduced immunity to pathogens and toxins, and in syndromes like the recently labelled Colony Collapse Disorder.

The Well-regulated Hive

Much has been written over the centuries esteeming the industrious well-

---

53 This is only possible because there is eros involved—that is, a relationship or sweetness.

54 Wilson, *The Hive,* p. 234.

55 Edwardes, *The Bee-Master of Warrilow,* p. 35.

regulated society of the hive as a model for humans.[56] The mystery of the bees and the honey, and the life of the whole hive, has traditionally been approached with respect and reverence, in some ways as a numinous secret, which if unlocked could show humans how to live ethical and sacred lives in a cooperative, orderly and productive way. Recently this hope was expressed when scientists embarked on the mapping of the honey bee genome. They hoped that this would allow them to glean new information on how humans can overcome some of our modern social, health and commercial organisational problems, and how we can be more altruistic and efficient in groups. Here is a sample of the language of scientists bidding, at the proposal stage, for the honey bee genome.

> ***Summary.*** Homo sapiens *is a highly social species and social interactions are critical determinants of human mental and physical health. We propose to sequence the genome of another highly social species, the honey bee,* Apis mellifera. *Though phylogenetically distant, honey bees live in societies that rival our own in complexity, internal cohesion, and success in dealing with the myriad challenges posed by social life, including those related to communication, aging, social dysfunction and infectious disease. A honey bee genome-sequencing project (HBGP) will benefit human health and medicine in diverse areas, including venom toxicology, allergic disease, mental illness, infectious disease, parasitology and gerontology. In addition, the HBGP will improve human nutrition by enabling enhanced pollination of food plants and accelerated delivery of hymenopteran parasitoids for biological control of pests. The HBGP will also improve honey bee sentinel function, providing enhanced capabilities for detection and location of chemical and biological agents of harm. Sequencing the genome of the honey bee, a beneficial, non-dipteran insect endowed with a small brain but cognitive sophistication, with complex social organization but amenable to molecular, genetic, neural, and ecological manipulation, will provide important tools and unique models to improve human health. When these benefits are balanced against the costs of sequencing a 270MB genome, the HBGP promises to provide a valuable and economical resource.*[57]

---

56 For example, Bernard Mandeville's poem 'The Grumbling Hive', p. 170.

57 *Proposal for the Sequencing of a New Target Genome: White Paper for a Honey Bee Genome Project.* The Honey Bee Genome Sequencing Consortium (2002).

Contrast this with the alchemical view expressed by Seneca in a letter of about 54 A.D.:

> *It is our duty to copy the bees, who fly from flower to flower to gather honey. They carry home their harvest and place it in the combs. We are not, however, quite clear if the juice which they take from the flowers is ready-made honey, or whether by mixing some substance with it, or by means of some peculiarity of their breath, they change the material into honey. Some say the bees are only capable of collecting honey, not of preparing it; for in India honey is found on reeds, which is either made by the dew of that climate or by the rich and sweet juices of the reeds themselves. It is said that our herbs possess this peculiarity also though in a less striking degree, and that the bees are born to collect this marvel for themselves. Others are of the opinion that the bees change what they take from the flowers into honey, and for this purpose use some sort of ferment which possesses the power of changing the different substances into one homogeneous mass.*[58]

Ultra-violet and Infra-red: the Art of Karl von Frisch

Jung used the simile of the colour spectrum to differentiate between the poles of the instinctual and archetypal. At the infra-red end are the instincts and physiological processes, and at the other end (the ultra-violet or spiritual pole) are the archetypes as instinctual images. He wrote:

> *The dynamism of instinct is lodged as it were in the infra-red part of the spectrum, whereas the instinctual image lies in the ultra-violet part. . . . The realisation and assimilation of instinct never take place at the red end, i.e. by absorption into the instinctual sphere, but only through the integration of the image which signifies and at the same time evokes the instinct, although in a form quite different from the one we meet on the biological level.*[59]

Scientists have discovered that bees are red-blind, while the more fiery wasps 'see red'—an expression in English for being very angry.

---

58 Seneca, *Letters,* 84, 3. Quoted in Hilda M. Ransome, *The Sacred Bee in Ancient Times and Folklore,* p. 87.

59 'On the Nature of the Psyche', *The Structure and Dynamics of the Psyche,* CW 8, par. 414.

The following statements from Karl von Frisch are apt given Jung's discussion of the spectrum that ranges from the more basic instinctual ultra-red to the more refined and spiritual ultra-violet:

> *Bees are able to perceive ultra-violet rays. Ultra-violet is a special colour for them, distinguishable from blue and all other colours. It is evident that the colours of flowers have been developed as an adaptation to the colour-sense of their visitors.*[60]

This was a significant breakthrough in natural science because 'it confirmed that the colour of flowers did have a biological meaning'.[61] Wasps are attracted to red flowers (instinctual wasps) and bees to the blue-purple flowers (spiritual bees). I've witnessed this in my own garden: bees love the lavender and rosemary flowers, and wasps favour the red poppies.

Von Frisch's research was based on a strong empathy with the bees—it was as if he were trying to feel (or in other words, actively imagine) what it was to be a bee. He discovered many marvellous and sometimes intricate details about the life of the bees, including for example that the bees' well-known liking for sweetness is fastidious. The sweetness of the bee itself is also legendary, as illustrated in the following beautiful story of a bee that carried crumbs from the Sultan's table to feed a poor blind sparrow. The poet Nizami (ca. 1200 A.D.) wrote:

> *A hornet seized a bee, intending to devour it. The bee begged for its life and said. "My hive contains so much honey, I myself am of so little worth, why do you not go to the hive and leave me in peace?" The hornet replied, "The hive is sweet because of the honey in it, but you must be sweeter as you are the source and spring of the honey."*[62]

The bees' customary nectar has a sugar content of forty per cent, which in honey making is intensified. Von Frisch's experiments also found that bees have a stronger sensitivity to salt than humans do, but that bees are less sensitive to bitterness. 'They [bees] seem to enjoy a

---

60 Von Frisch, *Bees*, p. 10.

61 Wilson, *The Hive,* p. 245.

62 Ransome, *The Sacred Bee,* p. 74.

mixture of quinine and sugar which is so disgusting to the human sense of taste that anyone would spit it out at once'.[63] This is another unlikely opposite to consider in the symbolism of bees: we also talk often of 'bitter-sweet' experiences in love.

On considering the bees, Marie-Louise von Franz wrote:

> *Bees have an incredible unconscious cooperation of the whole group, though we think of them as completely unconscious and as having only a sympathetic nervous system. Karl von Frisch described the most astonishing experiments made with bees. They can distinguish colours and can show each other where nectar is. A bee will fly in a big circle and find honey,*[64] *but does not have to fly back the same way; it can go straight and by certain movements of the back and wings can show and tell the others so that they can go straight to the honey. It has been discovered that the bee's system of orientation has to do with the polarisation of sunlight. The unconscious instinct of bees is amazing, and they therefore often symbolise harmonious functioning without rational organisation.*[65]

Von Frisch's research into the life of bees actually challenges the view that bees are unconscious. When dancing, bees are conveying information like human beings would. The scout bees (females), when foraging for nectar and honey sources, convey information to the foraging groups of worker bees in the hive, in dance language related to the sun and distance of the source from the hive. This enables those following to fly (in a 'beeline') straight to the source. In humans this would be regarded as a conscious and intentional act. These kinds of communications, as Jung emphasises, would only be regarded as unconscious in exceptional cases:

> *Nevertheless it would be possible to suppose that in bees this process is unconscious. But that would not help solve the problem, because we are still faced with the fact that the ganglionic system [in bees] apparently achieves the same result as our cerebral cortex. Nor is there any proof that bees are unconscious.*[66]

---

63 Crane, *A Book of Honey,* p. 25, n. 19.

64 It is not usually honey they are looking for, but nectar or pollen sources.

65 Von Franz, *Shadow and Evil in Fairy Tales,* pp. 52f.

66 'Synchronicity: an acausal connecting principle', *The Structure and Dynamics of the Psyche,* CW 8, p. 956.

Von Frisch's observations of bees and the hive strongly suggest the existence of transcerebral thought and perception. The honey bee genome has in fact revealed that the bees have links with vertebrates that no other insects have. Biologically and, as we shall see, symbolically, the bee bridges not only the realms of plant and animal life, but also those of insects and vertebrates—so much so that scientists expressed, when the honey bee genome was published, that they hoped it would help them improve agriculture and human health, including mental health, through bee-brain research. In fact they found that the bees' way of changing gene signals tallies with that which previously was thought to be found only in vertebrates. For example, Gene Robinson, at the University of Illinois at Urbana-Champaign has used microarrays to study this and has found that the expression of thousands of genes changes in the honey bee brain in response to social stimuli, like the hive's need for more foragers.

It seems to be a synchronicity that at the same time as the bee genome was released to the collective, what has been labelled Colony Collapse Disorder in bees also appeared as a news headline in the Western world. One example in the local press in 2006 was about Australian National University's Professor Mandyam Srinivasan's research focused on the bee brain. In the five years since 2001, for example, the US military and NASA awarded him research contracts totalling $4.5 million These contracts involved duplication of bee visual navigational skills to create low-cost, lightweight technology that could pilot unmanned aircraft on Earth or Mars. He says a bee brain has about a million neurons and humans have a million-fold more. But bees somehow manage to store complex maps of their world, achieve amazing navigational tasks and even pass on information to other bees. 'The big mystery is how their brain does it', he said. 'One clue is that the bees filter out a lot of irrelevant information.'[67] Bees dance to report where food can be found, can remember scents, navigate to a single leaf in a forest and even find deliberately camouflaged objects. But they are lost only if their hive is shifted a few metres while they are away. They will be disoriented and keep coming

---

[67] Richard Macey, 'The buzz in science is what makes a bee so smart', in *Sydney Morning Herald*, 17 October 2006, p. 6, in discussing Srinivasan's research.

back to the original precise location.[68]

The mystery of the bees continues to be numinous now as ever. Today one of the key questions in what is called insect socio-biology is: 'How does adaptive colonial behaviour emerge from the actions of individual colony members?'

Leading bee-research scientist Gene Robinson tells us that 'the honey bee system is in fact very flexible. [The bees] can accelerate, delay, or even reverse their maturation depending on the needs of the colony.'[69] How such changes are regulated was particularly interesting to him because social insects like the bees, he reminds us, operate under a complex system whereby a higher-level entity, the colony, makes decisions. This is conceivably comparable to the authority of the unconscious and the Self over the individual ego.

I will further explore this here later, applying Jung's psychological views and the idea of *zoë* and the alchemical 'mystery of the queen' as expressed through circumambulation. The quest will involve looking for symbols that, when amplified, can shed some light on what the bee, the honey and the hive might point to in our dreams and in more collective stories. Remembering that, as with the collective unconscious, the mystery and unknowing remain and, as Karl von Frisch reminds us, 'the life of the bee is like a magic well: the more you draw from it, the more there is to draw'.[70]

---

68 Ibid.

69 'Profile of Gene E. Robinson', in *Proceedings of the National Academy of Science* (PNAS), vol. 103, no. 44 (31 October 2006), pp. 16065f.

70 *The Dance Language.*

2
Bee Symbolism in Myth, Legend and Story

The earth bore much substance, on the mountain the oak at its top yielded acorns and midway, bees.[71]

Perhaps in this way we shall attain the high philosophical goal of perceiving how the divine life in man is joined in all innocence with animal life.[72]

Humans have been fascinated by and have treasured bees for millennia—venerated for their marvellous golden substance, yet feared for their sting. The way bees live, work and produce honey is by and large still a living mystery. The honey, the bees' food and raison d'etre, is not just something of concern to them; it is of paramount importance to humankind, and the bees as honey producers and pollinators are like the saviours of the macrocosm mentioned in alchemy.[73]

Honey is good for the health and came to be recognised very early in human history as a true giver of life, a substance necessary for existence, like salt, water and milk. The bee was held to be a creature of special sanctity connected with what served the sacred and numinous fundamental human mysteries of birth, death and reincarnation. The hive, the living community of bees, is a powerful example of natural vitality and self-regulation in one community of the spirit. An exploration and amplification of the symbolism of bees, the honey and the hive leads into concepts of balance in nature, of the Great Mother, and of the embodied spirit—that is, spirit embodied in matter or in the natural world.

****

71 Hesiod, *Works and Days,* 233.

72 This is the dedication quote by Goethe, in Kerényi, *Dionysos: Archetypal Image of Indestructible Life.*

73 See my section in chapter 4 on the Transformed Berserker vision.

The Bee Goddess and the Great Mother

Feminine Potency and Fertility

The magico-religious significance of bees is ancient. Marija Gimbutas has researched the myths, legends and cult images from 7000 to 3500 B.C., and her work shows that bees were counted among the sacred objects of this early period. Neolithic Northern Europe was what she calls 'matrifocal' rather than matriarchial. Later artefacts, for example, depict the bee in a way that connects it with the Great Goddess. She was depicted as 'Queen Bee' by the Minoans. Bees were believed to have been closely tied to bull worship which also used to be dedicated to the Great Mother Goddess. Figure 5 combines both bee and bull.

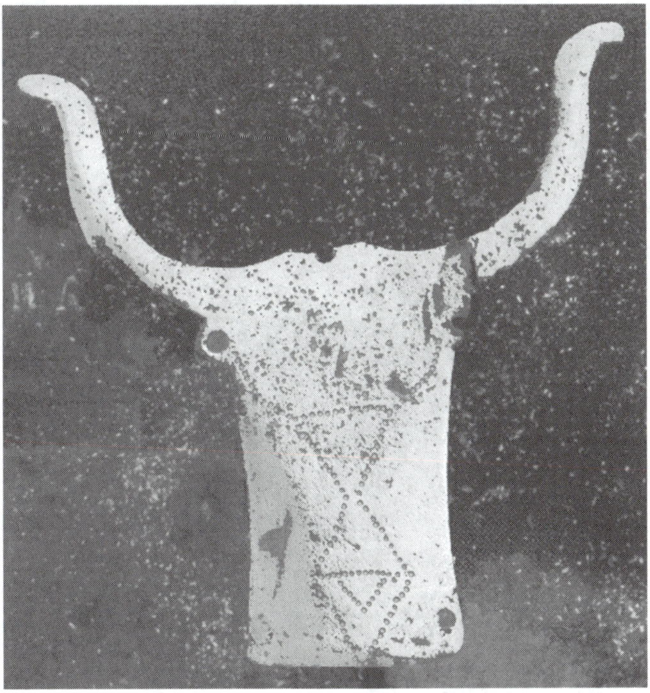

Figure 5. Goddess of regeneration: bee-form with antennae, 4000–3500 B.C. (Late Cucteni). This is on a stylised bull's head (17 cm high) carved from bone; Bilcze Zlote, Ukraine; (after photograph by M. Gimbutas).

Also of note are the bee-masked priestesses that appear on Minoan seals, and the goddess figure of Merope (meaning 'honey-faced') in Greek mythology. It is possible, then, that the female representation found in the pendant of gold bees (Figure 6) is not merely decorative, but an intentional composition created to symbolise the Great Goddess.

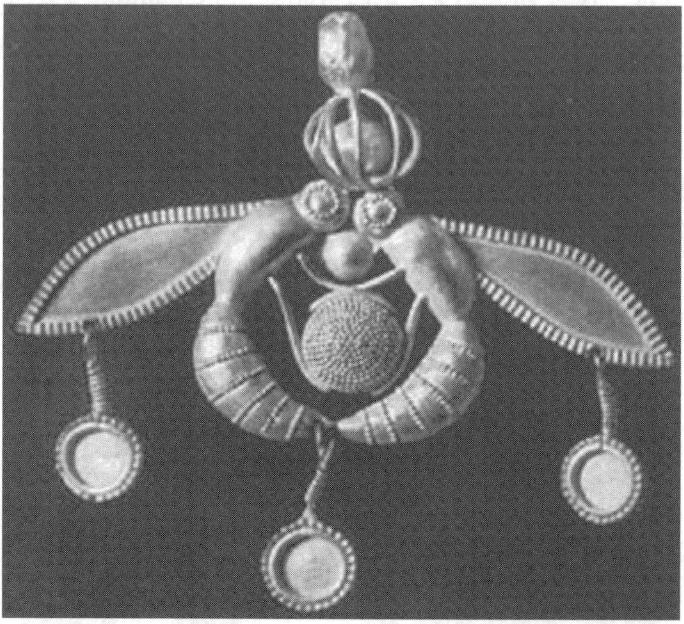

Figure 6. Visually this gold pendant of two bees (Malia, Crete 2000–1700 B.C.) when taken as a whole has the rounded shape of a woman. The bail of the pendant is like a head, while the large eyes of the bees become the breasts of the woman and their connected tails form her womb, containing a disc, perhaps representing the seed of life.

A bee figure was also drawn on the leg of a splendid Minoan offering table, dug up at Phaistos in Crete. This is from the Proto-Palatial period (2000–1700 B.C.). This figure, sometimes called a bee goddess, has a dewdrop form that suggests a dew or bee cult may have been a feature of Minoan religion and combines several other aspects of the goddess as well. The Medusa-like head is crowned with snakelike tresses, linking this figure with Medusa and Athena, and has the short, blunt beak of a dove. Heavily drawn arms ending in feathery fingers or claws contrast with the lightly drawn feet. Several other drawings in Arthur Evans's *The Palace of Knossos* and work by archaeologist Marija Gimbutas corroborate that the bee goddess had a central place in ancient Crete (Figure 7).[74]

Figure 7. Bee-headed goddess medallion, from Knossos, Crete, 1500 B.C. (Late Minoan II.) The goddess has two sets of bull horns and a double-axe-shaped butterfly. The winged dogs flanking her reinforce the theme of becoming.

Anne Baring and Jules Cashford, poetically and with eros, also explore the origins of the bee cult, honouring this dynamic of flux and flow so important to the feminine and to the archetype of the Great Mother:

> *Bees as well as all insects that spin cocoons or weave webs serve as images of the miraculous interconnectedness of life. The intricate intercellular structure that secretes the golden essence of life is an image of the network of invisible nature that relates all things to each other in an ordered harmonious pattern. Perhaps this is the meaning of the tale in which*

---

74 Buffie Johnson, *Lady of the Beasts: the Goddess and her Sacred Animals,* p. 157.

> *the infant Zeus is fed on honey in Crete [in a cave], and why honey was the nectar of the gods. Furthermore the busy bee, following the impulsion of its nature to pollinate the flowers and gather their nectar to be transformed into honey, was an example of the continual activity required of human beings to gather the crops and transform them into food. The queen bee, who all the others serve during their brief lives, was, in Neolithic times, considered to be an epiphany of the goddess herself. For a watchful eye, the relationship between the queen bee and the goddess must have seemed irresistible, and in Minoan Crete 4,000 years later the goddess and her priestesses, dressed as bees, are depicted dancing together on a golden seal found buried with the dead. The hive was her womb—perhaps also an image of the underworld—and later reappears in the beehive tombs of Mycenae.*[75]

The cave and the palace-temple in the Palaeolithic era were the main focus of Cretan religious life, with the labyrinth as the image of an initiatory path leading to the central mystery of the Minoan ritual. In Crete, the Great Goddess was experienced as a flowing, dynamic energy that could manifest in a swarm of bees, a dolphin's joyous leap, the flight of birds, the coiling of serpents and sea creatures as well as in the human gesture.[76] This energy is fundamental to the health of the hive as it is for the *ecclesia spiritualis* (understood alchemically and psychologically as those who take part in the alchemical *opus*—transformation and union of opposites).

On the boundary between the plant and animal realms, both governed by the Great Mother, lies the bee. In the oldest times, along with milk, the bees' honey was sacrificed to the earth goddesses. This provides a contrast to the bloody death symbolism of the animal world and the bee, a kind of intermediary between plant and animal, is a favourite with the Great Mother.

J. B. Bachofen, whose writing and studies focused on matriarchy, describes the bond between the beehive (in this case, the unity of the 'female' worker bees) and the one Queen Mother, in opposition to the many and 'alien' father drones, and summarises the symbolic significance of

---

75 Baring and Cashford, *The Myth of the Goddess*, p. 73.

76 Ibid., p. 107.

the bee and the hive as follows:

> *This makes the beehive a perfect prototype of the first human society, based on the gynocracy of motherhood . . . . Aristotle goes so far as to place the bees higher than the men of that first period, because in [the bees] the great law of nature is expressed far more perfectly and firmly than among men. The bee was rightly looked upon as a symbol of the feminine potency of nature. It [the bee] was associated above all with Demeter, Artemis, and Persephone. Here it symbolised the earth, its motherliness, its never-resting, artfully formative busy-ness, and reflected the Demetrian earth soul in its supreme purity. Its relationship with all physically conceived motherhood was expressed in a custom recorded by Heraclides in the Syracusan Thesmophoria, the participants carried mylloi, cakes made of honey and sesame in the shape of the female sex organ. In his monograph on bees Menzel draws an apt parallel between this custom and the Hindu custom of daubing the woman's genitals with honey at the marriage feast. In Germany the honey flower melissa [lemon or bee balm] is called Mutterkraut, mother weed, and is regarded as a specific [remedy] for feminine sexual ailments. The bees express their motherhood also as nurses and feed the newborn infant Zeus with honey. The purest product of organic nature, in which animal and vegetable production seem so intimately intertwined, [honey] is also the purest food for mothers. It was used by earliest mankind, and priestly men, the Pythagoreans, Melchizedek, John the Baptist reverted to it. Honey and milk belong to motherhood; wine to the male Dionysian principle.*[77]

The virginity of the Great Mother (that is, her independence of the male) becomes particularly evident in what can be called an Amazonian bee state, where only the queen is fecundated by the male (drone) and she goes on only one nuptial flight out of the hive (described below in chapter five). For this reason, and because of the food (royal jelly) she eats, the bee is considered as pure. Demeter is regarded as the pure mother bee and so her bee priestesses must be virgins, like the Vestals and many other priestesses of the Great Mother.

---

77 Bachofen, *Mutterrecht,* pp. 114f. The mythology of Dionysos is also linked with honey, as Kerényi mentions in *Dionysos,* and as Marcel Dietienne explores (see *Dionysos Slain,* chap. 3, p. 7).

Figure 8. The Goddess Spes with sheaves and beehive
(from a Roman original).

And among the bees, as so often among beasts and men, matriarchal womanhood assumes a character of the 'terrible' in its relation to the males: for after mating, the drone-mate [dies] and [then later] all the other drones are slain like aliens by the female.[78]

Marija Gimbutas in her extensive work on the goddess states that the bee and butterfly belong together as images of the Great Goddess of Regeneration. It was a very ancient belief that bees arose out of the dead carcass of a bull, and the association of bee and bull is made as early as Neolithic times in the image of the bee goddess incised on the head of a bull (as shown in Figure 5 at the start of this chapter). This symbolism remained and developed. What the ancient Greeks called *bugonia* ('birth from an ox') focused on the widespread belief that bees could spontaneously regenerate from the carcass of an ox or bull (Figure 9, next page).

The Latin poet Ovid wrote of this awe-striking regeneration: 'Swarms of bees rush from the putrid ox: one life extinguished created a thousand.'[79] He also wrote of recovering dead bees 'by art'. And Jules Michelet, a nineteenth-century French historian, called Virgil's fourth *Georgic* (which describes the bull-born bees created through the ritual process of *bugonia* in the context of the myth of Aristaeus) 'a song full of immortality, which, in the mystery of the transformations of nature, contains our dearest hope: that death is not death, but a new life beginning'.[80]

The belief in *bugonia,* the spontaneous regeneration of life, in particular bees from the corpse of an animal, continued for over 2,000 years.

---

78 Baring and Cashford, *Myth of the Goddess,* p. 107.

79 Ovid (43 B.C.–A.D. 18), *Fasti* I, 376ff.

80 J. Michelet, *L'Insecte*, p. 305.

Figure 9. *Bugonia:* 'bull-born' bees. Wenceslaus Hollar's illustration in Dryden's translation of Virgil's fourth *Georgic* (1697).

The Birth of Zeus: Sky and Cave, Light and Honey

The symbolism of the bee, the honey and the hive links the animal and plant realms with the divine in many ways in different lands and cultures. Various stories from ancient Crete emphasise the centrality of both the sky god Zeus and the Great Mother whose realm was the earth and extended to the underworld. The fruitful life force of Dionysos, like that of the bee (in particular the queen bee) and Mercurius in alchemy, bridges the two realms with eros and a passion for life.[81] This has the potential for flux, overflowing, as well as dismemberment and stinging self-sacrifice. The bee's sting protects the source of life's potency and fecundity, embodied in both the queen and the hive.

The opening quote of this chapter, the epigraph by Goethe in Kerényi's *Dionysos,* invites the reader to experience in a more enlivened, embodied way what underlies the Dionysian dance or rhythm of indestructible life. Its second chapter is called 'Light and Honey' and is about the overflowing of light and honey in connection with the birth of a divine child. For the Greeks this child was Zeus, whose name originally meant 'lighting up' but over time came to mean 'he who illumines'. Kerényi, focusing on the beginnings of a significant strand in myth, turns to the birth cave in ancient Crete and to a story called 'Thieves':

> *In Crete it is said there is a cave of bees. There according to the myth Rhea gave birth to Zeus, and a religious commandment* (hosion) *decrees that no one, neither god nor man, may enter. At a certain time of the year a great fiery glow is seen emerging from the cave. According to the myth, this is when the blood remaining from the birth of Zeus overflows. The cave is inhabited by sacred bees, the nurses of Zeus.*[82]

The story goes on to describe how four men tried to rob the honey from these sacred bees and how Zeus the supreme sky god, who was greatly angered by this violation, wished to punish them severely. As there could be no deaths in the cave, rather than being slain by Zeus's thunderbolt they were transformed into birds that, like bees, symbolise

---

81 That is, touched by the flow of the water of life.

82 Kerényi, *Dionysos,* pp. 30f.

the ascending and descending spirit.[83]

The idea of overflowing or 'streaming' of the light or blood, as divine life force, maintains the link with the bedrock dynamic of the chthonic Great Mother. Kerényi notes that 'for the Cretans the divine birth flared up from dark depths. It was more the bursting forth of life than the flaring up of a more spiritual, pure light.'[84] The sense of flow has not been lost here.

Dionysos

According to Ovid,[85] it was Dionysos who was the inventor of honey and his retinue, the maenads, were said to have danced on ground flowing with milk, wine and 'the nectar of bees'.[86] Milk flowed and honey gushed from natural springs wherever they went. These unadulterated natural substances are the divine and miraculous sources of nourishment in many descriptions of paradise or the Promised Land.

The Homeric *Hymn to Dionysos* celebrates the life-force associated with this god of the dance and clusters (like grapes and swarms of bees), of ecstasy and sexual passion in which heaven and earth are united in a transforming consummation such as the *hierosgamos* of alchemy or the nuptial flight of the queen bee:

> *I begin to sing of ivy-crowned Dionysos, the loud-crying god,*
> *splendid son of Zeus and glorious Semele.*
> *The rich-haired Nymphs received him in their bosoms from the lord*
> *his father*
> *and fostered and nurtured him carefully in the dells of Nysa,*
> *where by the will of his father he grew up in a sweet-smelling cave,*
> *being reckoned among the immortals.*
> *But when the goddesses had brought him up, a god oft hymned,*
> *then began he to wander continually through the woody coombes,*

---

83 The bee can also symbolise the winged volatility of the soul rising up and flying away from the body in shamanic traditions including those of Siberia, Central Asia and South America.

84 Kerényi, *Dionysos,* pp. 32f.

85 *Fasti,* III, 736.

86 Kerényi, *Dionysos,* p. 31.

thickly wreathed with ivy and laurel.
And the Nymphs followed in his train with him for their leader;
and the boundless forest was filled with their outcry.[87]

And so hail to you, Dionysos, god of abundant clusters!
Grant that we may come again rejoicing to this season,
and from that season onwards for many a year.[88]

****

The new life of both these great but very different gods, Zeus and Dionysos, was fed with honey in birth caves of sweetness. Wild bees often make their homes in dark caverns. In many traditions and rituals of mortals, the first taste for an infant is of honey and milk: the life-promoting victuals. Honey is the food of the bees. It is what they work for. It is their water of life. Bees, with their sweet food, were 'the most natural nurses for a Divine Child who was born and then kept hidden in a cave. The archetypal situation that nature offered was taken into the myth of Zeus. The bees offered men the essential sweetness of pure existence—the existence of infants in the womb.'[89]

This theme is connected in world mythology with the eternal return or the indestructibility of life. Honey is also a significant symbol in the language of deep eros and love. Its soothing sweetness is often present in the contexts of sensuality and the erotic. Honey, as in the Grimm's fairy tale 'The Queen Bee', has a double aspect. It is, on one hand, a soothing substance that can help induce sleep in much the same way as warm milk and honey is used as a calming, sleep-inducing nightcap; on the other hand, honey is also an elixir of life—renewing and regenerating vitality.

Aristaeus: God of Beekeeping and Bull-born Bees

To consider the Greek myth linked with founding of the bee-regeneration rite of *bugonia,* that of the 'bull-born bees', we will turn to the story of the god Aristaeus. One of the strands in the mythology of ancient Greece concerns Orpheus (son of Oeagrus, a river god, and—it is usually said—

---

87 *The Homeric Hymn to Dionysos,* XXVI, ll. 1–9.

88 Ibid., (ll. 10–12).

89 Kerényi, *Dionysos,* p. 31

of Calliope, chief of the nine Muses and muse of lyric poetry). The story has it that Orpheus could sing so sweetly that wild beasts would follow him, trees and plants would bow down to him, and the wildest of men would become gentle when they heard his singing.[90] The Orpheus cult has honey as a central motif.[91]

In considering the significance of this, the story of Aristaeus (or Aristaios)—a god closely linked with bees and beekeeping—is one starting point.[92] His myth contains examples of regeneration and *zoë:* the 'awakening of the bees'. Kerényi writes of Aristaeus that his greatest pride was his bees. His name signified that he was 'the best' that the world has 'to show' and that 'the 'honeyed' Zeus of the dead, Zeus Meilichios, who was accustomed to receive the worship of the living in serpent-form, was no other than Aristaios [Aristaeus]'.[93]

****

Aristaeus is described either as a son of Uranus and Gaia, or, according to a more general tradition, as the son of Apollo (the sun god) by Kyrene, a nymph (nature spirit) and the granddaughter of Peneius. Other, more local, traditions, say his father was Chiron and that Aristaeus was a firm favourite of the gods. His mother Kyrene had been spirited away from Mount Pelion (the place of Chiron's cave-home) and taken by Apollo to Libya where he found her boldly fighting a lion with her bare hands (like the Biblical Samson). She gave birth to Aristaeus in Africa. As the story goes, Pindar sang, when Aristaeus was born, that Hermes took the infant to be raised on nectar and ambrosia, the honeyed food of the gods.[94]

As a youth, Aristaeus went to Thebes in Boeotia, where he learned the

---

90 Pierre Grimal, *Dictionary of Classical Mythology,* p. 331.

91 See Marcel Deitienne's article, 'Orphée au miel'.

92 Aristeaus was worshipped in various parts of ancient Greece including Thessaly, Keos and Boeotia, but especially on the islands of the Aegean, Ionian and Adriatic seas. Different accounts of Aristaeus have arisen in different places and independently of one another so that, before a later merging, they referred to several distinct beings and initially to a mortal who then became a god.

93 Kerényi, *The Heroes of the Greeks,* p. 282.

94 Pindar (c.522-443 B.C.), *Pythian,* ix, 112f.

arts of healing and prophecy from Chiron and the Muses.[95] After the unfortunate death of one of his sons, Actaeon, Aristaeus left Thebes and went to Keos, whose inhabitants he delivered from a severe drought by erecting an altar to Zeus Icmaios ('rainmaking Zeus'), the god of moisture, who sent the cooling Etesian winds. Honey was used as the libation. His other child Macris nursed the child Dionysos. Aristaeus returned to Libya, where his mother prepared a fleet for him. He sailed to Sicily, visited several islands of the Mediterranean, and for a time ruled over Sardinia. His cult spread throughout Greece and its colonies. Finally he went to Thrace (Asia Minor), where he was initiated into the mysteries of Dionysos. After dwelling for some time near Mount Haemus, where he founded the town of Aristaeon, he disappeared.

F. Hiller von Gaertringen says of Aristaeus that he has been characterised as 'a figure in Greek religious history, concerning whom we have only a fragmentary and dispersed tradition. From this tradition, however, the former importance of this god can be gathered. He was said to have imparted the use of the beehive to man and was looked upon as the inventor of mead.'[96] He belongs to a period of Mediterranean culture when instead of robbing hives in the wild, humans could tend their own beehives. He was one of the divinities most beneficent to mortals in ancient mythology. He was worshipped as the protector of flocks and shepherds (with the epithet 'ever close follower of the flocks', and also known as 'the pastoral Apollo'). Aristaeus was the protector of vine and olive plantations; he taught men to hunt and keep bees, averted the burning heat of the sun from the fields, and protected them from other causes of destruction. He was a *theos nomios* (protector of pastures and shepherds), *agreus* (hunter), and *alexêtêr* (healer). According to Virgil, the benefits Aristaeus conferred upon humankind varied in different places according to their particular needs. For example, to hot and drought-prone Keos he brought rain and refreshing winds, and in Thessaly and Arcadia he was the protector of the flocks and bees.[97]

---

95 According to Hesiod, Aristeaus married Autonoë, the daughter of Cadmus, who bore him several sons: Charmus, Calaicarpus, Actaeon and Polydorus. (*Theogony*, 975.)

96 Von Gaertringen, *Aristaios*. In Cook, *Zeus:A Study in Ancient Religion,* p. 267.

97 See Virgil, *Georgics,* IV, 283, 317, and *The Chiron Dictionary of Greek and Roman*

As a pastoral deity, Aristaeus embodied the qualities of both Apollo and the nymph Kyrene. Some mythographers say that, before nymphs were included in the stories, bees and goats were mentioned directly. (Pan certainly is also regarded as a god of bees and beekeeping, as is the phallic god Priapus.)

The other main version of the myth of Aristaeus has him as the son of Uranus, the supreme sky or father god (he was father of the Titans) and 'personification of the sky as a fertile element' that covers the whole of the Earth.[98] Uranus was the son–husband of Gaia, the Earth. And it was Gaia (Mother Earth) who made Aristaeus immortal. He was regarded as the patron of the hunt, agriculture, cattle, and especially bee- or api-culture. Aristaeus also taught humankind the very useful art of cultivating olives and the hunting skills of using nets and snares.

In the first version (Boetian) of the myth, when Aristaeus was born, Apollo, through Hermes, placed the infant in a cave in the care of his grandmother Gaia and the Horae —goddesses of the seasons and the powers of nature: Thallo (goddess of blossoms), Auxo (goddess of growth) and Karpo (goddess of ripe fruit). According to another tradition (more Arcadian), Aristaeus was brought up and educated by Chiron, the wise centaur. The Muses then added to Aristaeus's education by teaching him the arts of medicine and divination, which are strongly linked with honey. Aristaeus tended their flocks of sheep. This version emphasises the nymphs, the feminine nature spirits, as his teachers in agriculture, cheese making, farming and beekeeping, as well as in cultivating the grape vine and olive groves. Aristaeus was thus taught by the wisdom of Heaven (Uranus) and the Earth (Gaia), the instinctive wisdom and healing of Chiron, as well as by the cycle of the seasons. In turn, Aristaeus taught these arts and this practical agriculture to humankind.[99] He was described as being generous and beneficent to humankind.

Aristaeus pursued Eurydice, wife of Orpheus—a snake bit her and she died. Then Aristaeus's bees became ill and died. All his knowledge could

---

Mythology, p. 33.

98 Von Gaertringen, *Aristaios.* See Cook, *Zeus,* p. 267.

99 Grimal, *Dictionary,* p. 60.

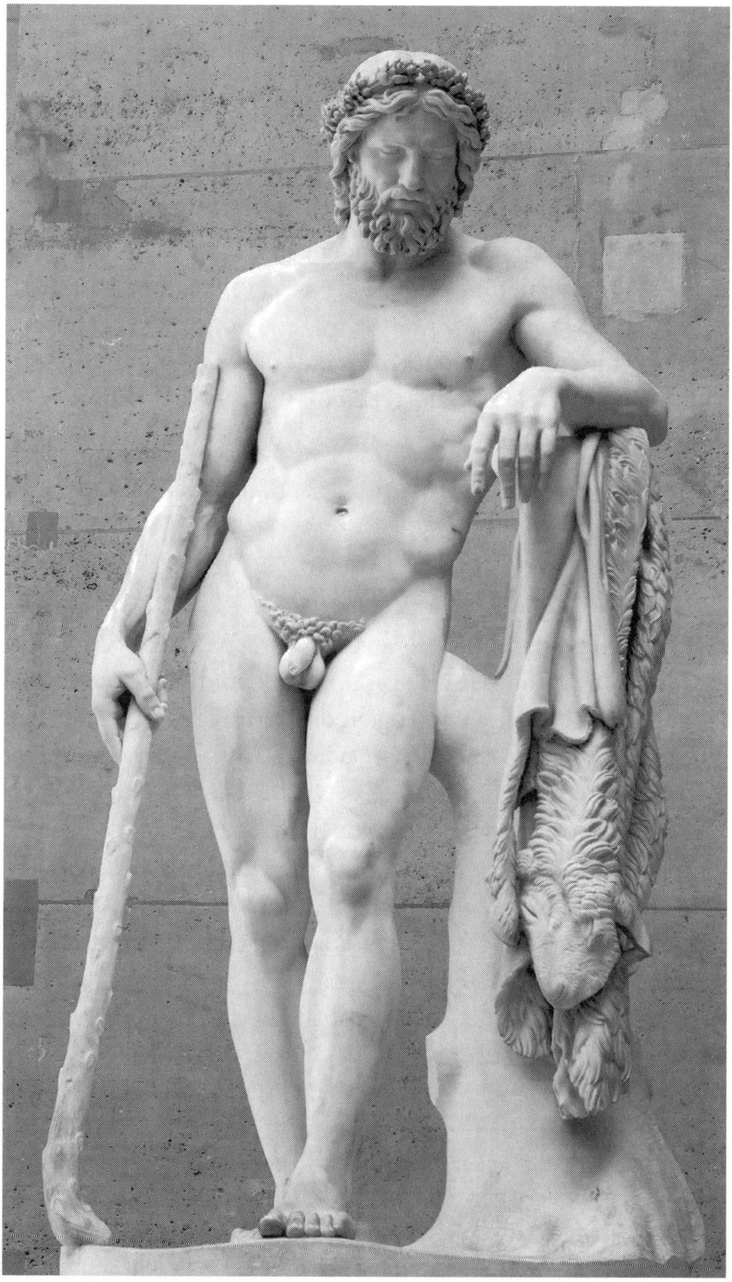

Figure 10. *Aristée, dieu des Jardins* (Aristaeus, God of the Gardens), by François-Joseph Bosio (1768-1845); Louvre, Paris.

not save his bees.[100] He had wronged a nymph and for a cure he had to turn to another nymph or feminine nature spirit (the wisdom of the anima and *anima mundi).* In despair he called on his mother Kyrene, who dwelt in a crystal palace beneath the waters of the Peneus. She told him that the only person who could help him was the sea-god Proteus (the wise but wily old man of the sea). Aristaeus went to question Proteus and found him sleeping near the mouth of the Nile on a rock, surrounded by a herd of seals, which he was looking after for Poseidon.[101] Proteus, a shape-shifter extraordinaire, in general did not like being questioned; so in order to ask him a question and receive an answer, Aristaeus had to tie Proteus up—to 'hold him fast'. In Ovid's words:

> *Thus captured he [Proteus] was forced to speak and told Aristaeus that he must bury the carcase of a slaughtered ox, and that from it he would obtain what he wanted, for when the carcass decayed, swarms of bees would issue from it. The death of one thus produced a thousand lives.*[102]

Proteus told Aristaeus that the gods were punishing him for Eurydice's death and gave him advice on how he could get some new swarms of bees. This involved the above sacred ritual of regeneration that came to be called *bugonia,* two versions of which Virgil describes.

Virgil wrote too of bees possessing 'a particle of divine intelligence',[103] and Porphyry says that the bees born from the carcass of a ritually prepared bull, as in the bugonia of Aristaeus, are souls that have incarnated or entered the world of creation.[104] This ritual preparation, which enabled the bees to regenerate spontaneously from a bull's carcass, is said to have originated in Egypt.

The following rather cruel but ritual method, Virgil says, was always practised in ancient Egypt:

---

100 As is the case with the scientists since 2006 and the outbreak of Colony Collapse Disorder affecting the bees in parts of Europe and North America.

101 See Michael Maier's alchemical *Peregrinatio* text and the quest for Mercurius; also Edward F. Edinger's *The Mysterium Lectures*, p. 150.

102 Ovid, *Fasti*, I, 363f.

103 Virgil, *Georgics* IV, 220f.

104 Porphyry, *The Cavern of the Nymphs,* 18.

> *Choose a small confined space, and erect in it a building with four windows, one facing each quarter, and with a tiled roof. Then take a bullock, whose second year's horns are just curling over its brow, stop up its nostrils and mouth and beat it to death without breaking the skin. Shut the bruised body up in the closed room, strewn with thyme and cassia, and after nine days the softened bones will have fermented, wondrous creatures will appear, who with buzzing wings will fly into the air—a swarm of bees.*[105]

No blood was to be shed in this process as it was from the blood that new life, the clusters of golden bees, would be born. From the strength of the ox or bull (often four bulls were specified in Virgil's version of the Aristaeus myth) would come the bees. A transformation, like fermentation or a bursting forth of life, takes place in a closed vessel or space (the sealed skin) and results in a miracle of regeneration. In another set of instructions, Cassianus Bassus says that 'a cubical structure should be built with a door and three windows, each turned towards one of the four cardinal points. . . . The animal is transformed into a sack containing its own fluids'.[106] Kerényi adds:

> *After four weeks and ten days—roughly forty days, as in the traditional brewing of mead—grapelike clusters of bees fill the hut. All that remains of the bovine is the horns, the bones, and the skin. The natural phenomenon [of the overflowing of the preserved divine blood or the light and honey from the birth cave of Zeus] ushering in a great festival for the early rising of Sirius, an ancient New Year's festival, was raised to the level of a myth of zoë: an awakening of bees from a dead animal.*[107]

Kerényi also describes a similar process from ancient Crete of regeneration of the life force (involving the fermenting of honey into mead). This too takes place in a closed vessel—a leather sack. Another version of this miracle, like that of the culminating process of *multiplicatio* and *projectio* in the alchemical *opus,* is found in the Old Testament. Here the bees and honeycomb spring to life out of the carcass of the lion that

---

105 Ovid, *Fasti,* I, 363f.

106 Kerényi, *Dionysos,* pp. 32, 40.

107 Ibid., p. 41.

Samson kills with his bare hands. I will discuss *multiplicatio* later in chapter five.

The Bee and the Muses: Inspiration and Divination

Bees abound in the mythology of many peoples and times. They have long been sacred to the ancient Great Mother Goddess and to the Triple Goddess, incorporating the maiden, mother and crone aspects of the feminine. Bees are also of primary importance in many mythologies and cosmologies. As mentioned earlier, in ancient Greece, after the birth of Zeus in a cave, he was fed with honey. Bees are also linked with Apollo, the god associated with light and consciousness.

There are some connecting threads in the various bee stories, like those of a relationship with the sun and the earth, and the Muses in various forms.

Bees are very closely linked with wisdom from both the heights and depths; they are winged messengers of the gods but they are equally at home in the dark hollows or caves of the earth. They are also Mercurial, as we shall soon see. The *Homeric Hymn to the Muses and Apollo* acknowledges the great role of the Muses in inspiring music and words:

> *I will begin with the Muses and Apollo and Zeus.*
> *For it is through the Muses and Apollo that there are singers upon the earth and players upon the lyre;*
> *but kings are from Zeus.*
> *Happy is he whom the Muses love: sweet flows speech from his lips.*[108]

Because they are also regarded as a source of celestial or divine inspiration, bees reputedly graced with honey the mouths of specific writers like Homer, Sappho, Pindar and Plato. Others, like Pythagoras and Saint Ambrose (an early Christian patron saint of bees), were also granted the gift of eloquence, as poets or investigators of the science underlying life. A bee resting on a living individual's lips also indicated the divine *pneuma* (spirit) returning to the body.

---

[108] *Homeric Hymn* XXV, II, 1–5.

Delphi and Divination: The Omphalos *as World Navel and Oracular Centre*

The oracle at the temple of Apollo at Delphi sat over a chasm. Such places invoke the chthonic energy of the Great Mother and wisdom of the snake, or of the depths and the unconscious. The Great Mother was known as the Queen Bee and her priestesses were called Melissae, the bees. Pindar says that the Pythian oracle at Delphi was known as the Delphic Bee, and her bee symbol was depicted on Delphic coins. These priestesses were like a retinue of worker bees circumambulating around their centre—the queen—ordering and regulating the temple or hive in serving the royal principle. As the queen bee is the only egg-laying fertile female in the hive, she is regarded as a symbol of feminine potency. Other instances of this bee symbolism in connection with the Mother Goddess and temple priestesses include the traveller Porphyry's description, from the third century A.D., of later goddesses of Greece:

> *The ancients gave the name of* Melissae *('bees') to the priestesses of Demeter who were initiates of the chthonian goddess; the name* Melitodes *to Kore herself: the moon (Artemis) too, whose province it was to bring to the birth, they called Melissa because the moon being a bull and its ascension the bull, bees are begotten of bulls, and souls that pass to the earth are bull-begotten.*[109]

The link between bulls and bees is expressed again here and was noted earlier in this chapter. In astrology, Taurus (the sign of the bull) is ruled by the planet Venus. Venus/Aphrodite, the goddess of love and mother of Eros, has much to do with the bee and honey. At the temple of Aphrodite at Eryx, priestesses were called *Melissae,* and Aphrodite herself was called Melissa, the queen bee. At the Ephesian temple of Artemis, the Melissae were accompanied by transgendered priests called *essenes,* which means drones (as the males in the beehive). In zoology, bees are classified as members of the *hymenopteran* order, meaning 'veil-winged', recalling the hymen or veil that covered the inner shrine of the Goddess's temple, and the high priestess who bore the title of Hymen.[110]

---

[109] Baring and Cashford, *Myth of the Goddess,* p. 118.

[110] S. Buxton, *The Shamanic Way of the Bee: Ancient Wisdom and Healing Practice of the Bee Masters,* p. 98.

Goddess Melissa
priestess Melissa

In Roman mythology the bee is sacred to Artemis and Diana; priestesses serving in the temples of these goddesses were also called *Melissae.*

So the links between the priestesses and the bees, as well as the bull, are many. To name these women, who had an important daily role in the sacred or religious life of a people, 'bee' means that the bees' work and life itself was regarded as extraordinary and numinous and was honoured as such in human culture. In Greece, Apollo's main temple and oracle (the Pythia) were based at Delphi near Mount Parnassus. Here, the spot reputedly marking the centre of the Earth was called the *omphalos* or navel of the Earth. This is the point where the two golden eagles that were released by Zeus to fly around the world met again, thus indicating the world's navel. It was here that the oracle sat on a tripod near a spot where vapours rose up through a chasm from the heart of the earth and the Castalian spring.

An *omphalos,* in general, is a numinous point in the centre sometimes marked by beautifully carved dome-shaped artefacts. A single significant stone often marks a sacred place or centre. Such a tall upright stone would be a kind of *axis mundi* linking heaven and earth. And the psychological dynamics of ascent and descent, as well as of circumambulation of a numinous centre, are present. These stones were often carved and shaped like skeps, traditional dome-shaped beehives, and some were decorated with carvings of bees with wings outstretched, as shown in Figure 11, to form a net covering the sacred centre.

These bees, with their veil-like wings spread open, form a net—one of the symbols of the Great Goddess. According to some reports like that of Philostratus the Elder,[111] the second temple of Apollo at Delphi was constructed of beeswax and feathers.

Omphalos stones are frequently viewed in depth psychological terms as images of the Self (because they are complete, that is unchanging—and lasting).[112] The *omphalos* or Delphic navel of the world was regarded as the durable, reliable and indestructible centre, out from which the whole cosmos radiated, and to which it referred back for stability and

---

111 *Vita Apollonius Tyanaeus,* vi. II. 4.

112 Jung, *Man and his Symbols,* p. 206. For example, there is a Hindu tradition of passing down, from father to son, stones believed to have significant esoteric power.

movement—like the bees leaving the hive to fly to the flowers at sunrise and ever-returning to their queen and community. The word *melissa,* as well as meaning bee, means priestess, poetry and healing.

Figure 11. *Omphalos* stone with net of bees with outstretched wings, Delphi.

The individuation process involves the symbolism of many sparks radiating from the centre, like the *lumen naturae* or the Egyptian sun god Re's tears, which became the bees. This symbolism also encompasses the alchemical *multiplicatio*—the effects of the living Stone, which emanates a radiating, multiplying effect from its essence. There is also the community of the Self to consider and this is symbolised in the life of the bee by the collective of bees, which together form the unity called the hive and which circumambulate this centre.

The Queen and 'King' Bee as Numinous Centre

Because the existence of the queen bee is central to the functioning of a healthy working hive, she, and what she serves, is the focal point of all activities in the hive. The sacred place of the oracle was similarly a focal point, like the *lapis* in alchemy and the inner-god image (the Self) in psychological terms. The queen bee as a central numinous point of focus emanates much energy, as in the process of *multiplicatio.* In ancient Egypt, where the bees were considered to be Re's tears, their movement described a golden multiplicity moving between heaven and earth and, in this case, also descending as moisture to fertilise the land.[113]

Neumann delineates the development of the theme of the great individual (which is so connected to Jung's formulation of the Self) through the queen bee, initially thought to be a king bee, but originally a god:

> *Historically the Great Individual is most accessible to us in the role of the god–king and, later, the king. The earliest cuneiform pictogram for 'king' signifies 'great man' . . . . The Great King or the Great House, Pharaoh, is the embodiment and representative of the people. If the hieroglyph for the king of Lower Egypt is the bee, and the same image occurs in the cultural sphere of the Euphrates, that only tells us the same thing. The 'great'*

---

113 The royal names of the early dynasties of Lower Egypt preferred the additional name 'Bee Prince'. In later times, the pharaohs received the title *n(y)-swt-bit* (He of the Sedge [Reed], of Upper Egypt, or He of the Bee, of Lower Egypt, representing the Nile delta area and as far up the Nile as Saqqara). The kingdoms of Upper and Lower Egypt were united around 3100 B.C. and from then (about 300 B.C.) the titulary of the king of Egypt, the Pharaoh, included the hieroglyph *nswt-bit.* Beekeeping was already very important, especially in the Nile delta, before 3000 B.C.

bee which rules the hive and which today we call the queen bee, was regarded in antiquity as the king bee. But in Egypt the designation of the king as the 'First Man' or 'Great Man' is already a later development. It follows the stage of his god-identity, when even ritually he was as far removed from his people as a god. Speaking of this stage, the Pyramid Texts say the king was already in existence before the creation of the world, an ideology which reappears later in connection with the Messiah.[114]

Significantly, the ancient Egyptians considered god as the 'Eye of the Universe', and a point within a circle was regarded by them as a symbol of the deity surrounded by eternity—the sun god Re.[115] They also regarded the eye as the seat of the soul. In alchemy the eye is heaven (the *coelum* or *caelum)*: 'It is like an eye and a seeing of the soul, whereby the state of the soul and her intentions are ofttimes made known to us, and through the rays and the glance [of heaven] all things take form.'[116]

Archaeologist Brian Hobley writes:

To the early Greek Pythagorean philosophers, the sun was identified as the heart/centre of the cosmos, the source of light and life. To Plato the sun represented the One as the source of existence. Apollo to the Pythagoreans was called 'the One'. . . . In the Greek alphabet, a circle divided into four quarters equated to theos, *God expressed as a point within a circle.'*[117]

Hermes was also known as 'The Eye of Heaven'—the fertilising centre. In dreams, the Self often manifests as a circle, a round thing, or through the image of the circumambulation of a numinous central point. All these symbols refer to 'an essential core of meaning',[118] to the circumambulation of the mystery of the soul.[119] This is the symbolism of

---

114 Neumann, *The Origins and History of Consciousness*, p. 428. See also p. 428, n. 3. In considering why the Pharaoh took the title 'Bee Prince', consider also the work of pyramid building, of the focused toil of thousands of humans for decades building the god–king monuments in order to assist the Pharaoh's self-immortalisation.

115 Brian Hobley, 'The Point in the Circle: The Image of Apollo in the Greco-Roman World', p. 32.

116 *Mysterium Coniunctionis,* CW 14, par. 46.

117 Hobley, 'The Point in the Circle', p. 32.

118 Von Franz, *Hall Ar-Rumuz (Clearing of the Enigmas),* p. 61.

119 A. Schweizer, 'The Teaching of Komarios and Cleopatra'.

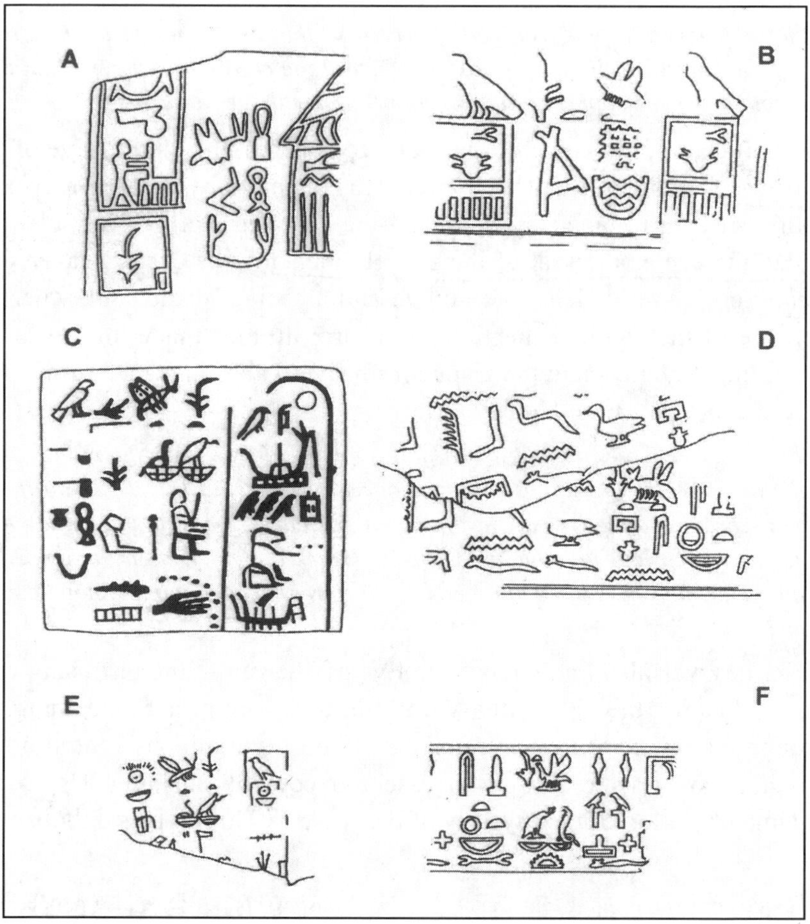

Figure 12. Egyptian bee hieroglyphics. Representations of the bee and the sedge in the title of early kings of Egypt: A–C, First Dynasty kings (c. 3100–2890 B.C.)—Den, Adjib, Semerkhet. D–F, Second Dynasty kings (ca. 2890–2686 B.C.)—Peribsen, Peribsen, Khasekhemwy.

the centre from which healing energy and illuminating consciousness or life force radiates or pulses out, just as bees do from the beehive.

Before returning to Hermes/Mercurius, I will first mention the myth of a swarm of bees and the founding of another oracle, and then briefly discuss honey as a ritual food.

Divine Inspiration: Oracle and Muse

Trophonius: The Rainmaker and the Swarm

In the mythology of various cultures, honey is associated with divination or prophecy. Pausanias tells the following story of the oracle of Trophonius, one of Apollo's sons who as an infant was suckled by Demeter:

Figure 13. Trophonius with beehive, fruits and flowers; from *Historia Deorum Fatidicorum* (Geneva, 1675).

This oracle, formerly unknown to the Boeotians, was discovered on the following occasion. No rain had fallen for more than a year, so they despatched envoys to Delphi from every city. When asked for a remedy for the drought, the Pythian priestess bade them go to Trophonius at Lebadea, and get the cure from him. But when they came to Lebadea and could not find the oracle, Saon, the oldest of the envoys, saw a swarm of bees, and he advised them that they should follow the bees wherever they went. Straightaway he observed the bees flying into the earth and followed them to the oracle.[120]

The bees flew into the earth and led others to the sacred site of earth-inspired wisdom. Trophonius was a rainmaker, like the bees in some myths and folk stories (see examples later).

As part of the incubation rituals associated with Trophonius, the initiate also descended into the earth—the cave—and crept, feet first, through a narrow tunnel. While doing this, he held honey cakes (or pieces of dough made of fine flour mixed with honey and poppy seed). After this, the incubant then heard or saw his oracle. Such honey-based ritual food was a typical offering to serpents in the cults of the chthonic deities including Cecrops and Erechtheus, who were worshipped in the form of serpents and given such honey cakes.[121] The sacred serpents in the Asclepia were also fed with these cakes, signifying that the snakes represented the chthonic aspects of Asclepius, the god of healing.

C. A. Meier explores the following connections: that honey cakes were offerings made to the chthonii; that the chthonii were prophetic: therefore, there is a link between honey and prophecy. The other links of honey and bees with the Delphic oracle have already been mentioned—namely that the priestesses dedicated to the oracle at Delphi were called Melissae (bees) and that Philostratus noted that the temple of Apollo at Delphi was constructed of beeswax and feathers.[122] This is a reminder that both birds and bees, in their ascents and descents, link the earth and

---

120 Pausanias, *The Description of Greece,* IX, 40. L.

121 Maier, *Healing Dream and Ritual,* p. 91.

122 Ibid., p. 92.

the realm of matter with the sky and the realm of spirit.

Mercurius/Hermes and the Bee Maidens

Bees are linked with the form of inspiration and communication called divination. The *Homeric Hymn to Hermes* tells of Apollo's gift to his newborn trickster brother Hermes of three bee maidens who drank mead and ate honey and (under the influence of the fermented, sacred honey drink) proclaimed their oracular wisdom by speaking the truth. If they were not fed honey, which the Hindus call the life or the sap of the sun (an image that evokes both the life urge and flow), the truth or *veritas* was absent and was not spoken.

These maidens represented something other than the oracle of Apollo the sun god.[123] Apollo's kind of wisdom took the form of logos-based, solar, far-shooting insights (like his golden arrows) and knowledge, whereas the white-faced bee maidens divined in a less straightforward way. They were Mercurial, embodying knowledge of a more flowing nature than only solar insights. The knowledge or wisdom of the bee maidens came to consciousness through a process in which the solar and lunar have been combined. The bee maidens were three, and Hermes/Mercurius was the fourth. The wisdom of the earth and the unconscious was present in the dynamic of this oracle—the gift of

123 Jungian analyst and astrologer Liz Greene notes that Apollo was the carrier of the sun or vessel of the ineffable, that which cannot be spoken or expressed: 'Just as the sun was perceived in the ancient world as the giver of life, Apollo as the representative of the sun was perceived as the giver of inner light. "Know thyself" was the dictum carved in stone at his shrine at Delphi, and this emphasises the importance of Apollo as a symbol of consciousness'. (*Apollon: The Journal of Psychological Astrology,* no. 1, p. 6) When considering our individual solar energy or life force, for example in a natal horoscope, one needs to consider being as well as doing: including the inner and spiritual levels of existence, a being quality like radiance or vitality. When you are 'on beam', life is purposeful and you feel vital and centred, almost regardless of what is going on in the external world. This is like being 'in alignment with the Self' in depth psychological terms, or 'in tao'. Even in the midst of emotional difficulties or external difficulties, there is an inner source of radiance or illumination that you can rely on as a guiding principle. This offers a sense of the life force of vitality: what it is like to be in touch with your individual purpose, your solar principle, your inner centre. This contrasts strongly with the opposite feeling of being 'endarkened'. It is to a sun god like Apollo, the 'far-shooter', that people would pray when they needed clarity (clear sight), for his arrows penetrate even the murkiest of dilemmas, and his music stills the confused and turbulent of heart.

conscious was present in the dynamic of this oracle—the gift of Hermes from the Sun God. That there is a significant link between honey and prophecy or divination is clearly stated, and also in connection with Delphi, in the *Homeric Hymn to Hermes* where Apollo gives his younger brother Hermes the gift of three bee maidens:

> *From their home they fly now here, now there, feeding on honey-comb and bringing all things to pass. And when they are inspired through eating yellow honey, they are willing to speak truth; but if they be deprived of the gods' sweet food, then they speak falsely, as they swarm in and out together.*[124]

These bee maidens are the three prophetic nymphs called Thriae. They lived on the sacred mountain of Parnassus at Delphi and taught the young Apollo the art of prophecy. These nymphs fed on honeycomb and after eating honey they spoke the truth. If they did not receive an offering of honey to eat, then, we are told, they spoke lies and just buzzed about chaotically together. It was also important to approach them with the right attitude, like that of a seeker or pilgrim, an attitude that would not create hubris or draw the wrath of the sky gods.

The white meal on the bee maidens is also linked with the blood sacrifice of the child Dionysos by the Titans with gypsum (white quicklime powder) smeared faces. Marcel Detienne says that Dionysos multiplies disorder, as when the bees are swarming, leaving their hive or when the bee maidens were not fed the sacred honey.[125] The maenads swarmed and danced around this god, but the creative side of this 'outbreak' was that married women were able to leave their homes at will and dance and 'flow' again through the wilds of nature, with springs of honey welling up where they trod. To dance with Dionysos meant to move between the domestic and the wild, men and beasts, gods and mortals. The darker more destructive side of this swarm erupts only when the god is not recognised. With bees, the swarm heralds that the queen and the life force of the hive needs renewal. Some radical change or rupture allows something new to well up: a new spring or balsam.

---

124 Hugh G. Evelyn-White, trans., *Homeric Hymn to Hermes* (ll. 550–568),

125 Detienne, *Dionysos Slain*, p. 54.

****

The Thriae, like the golden aegis, were special gifts from Apollo to his brother Hermes, part of the sun god's contract with Hermes/Mercurius. Zeus the supreme sky god sent out his golden eagle to confirm this negotiation. In alchemy, Mercurius is present at the beginning and end of the work. He is the *prima materia* and the *nigredo;* as a dragon, he devours himself and dies, resurrecting again as the *lapis* at the end of the work.[126] The alchemists reiterate that the *opus* is *circulare* (circular) or else *rota* (like the rotation of a wheel). Mercurius becomes the inner guide and way marker. Jung says:

> *In a psychological sense Mercurius represents the unconscious, for this is to all appearances that 'spirit' which comes closest to organic matter and has all the paradoxical qualities attributed to Mercurius. In the unconscious are hidden those "sparks of light"* (scintillae), *the archetypes, from which a higher meaning can be extracted. The magnet that attracts the hidden thing is the Self.*[127]

The following extract from the *Homeric Hymn to Hermes*[128] tells how Hermes/Mercurius [Maia's son], carrier of the three-branched golden aegis or wand, was given certain gifts of healing and how his particular gifts of divination came through the honey and the three bee maidens who flew around. The number three indicates dynamism. Honey or eros inspires the feminine, and the truth of the heart is in the wisdom bridging the unconscious and the conscious that Mercurius can teach and communicate to humankind. The importance of the honey here is great.

> (ll. 521–549*) Then Maia's son nodded his head and promised that he would never steal anything of all the Far-shooter possessed, and would never go near his strong house; but Apollo, son of Leto, swore to be fellow and friend to Hermes, vowing that he would love no other among the immortals, neither god nor man sprung from Zeus, better than Hermes: and the Father sent forth an eagle in confirmation. And Apollo swore also: 'Verily I will*

[126] *Psychology and Alchemy,* CW 12, par. 404.

[127] *Mysterium Coniunctionis,* CW 14, par. 700.

[128] In Evelyn-White, trans., *Homeric Hymn to Hermes.*

make you only to be an omen for the immortals and all alike, trusted and honoured by my heart. Moreover, I will give you a splendid staff of riches and wealth: it is of gold, with three branches, and will keep you scatheless, accomplishing every task, whether of words or deeds that are good, which I claim to know through the utterance of Zeus. But as for sooth-saying, noble, heaven-born child, of which you ask, it is not lawful for you to learn it, nor for any other of the deathless gods: only the mind of Zeus knows that. I am pledged and have vowed and sworn a strong oath that no other of the eternal gods save I should know the wise-hearted counsel of Zeus. And do not you, my brother, bearer of the golden wand, bid me tell those decrees which all-seeing Zeus intends. As for men, I will harm one and profit another, sorely perplexing the tribes of unenviable men. Whosoever shall come guided by the call and flight of birds of sure omen, that man shall have advantage through my voice, and I will not deceive him. But whoso shall trust to idly chattering birds and shall seek to invoke my prophetic art contrary to my will, and to understand more than the eternal gods, I declare that he shall come on an idle journey; yet his gifts I would take.

[Apollo, talking to his brother Hermes, then says:]

(ll. 550–568) *'But I will tell you another thing, Son of all-glorious Maia and Zeus who holds the aegis, luck-bringing genius of the gods. There are certain holy ones, sisters born—three virgins gifted with wings: their heads are besprinkled with white meal, and they dwell under a ridge of Parnassus. These are teachers of divination apart from me, the art which I practised while yet a boy following herds, though my father paid no heed to it. From their home they fly, now here, now there, feeding on honeycomb and bringing all things to pass. And when they are inspired through eating yellow honey, they are willing to speak truth; but if they be deprived of the gods' sweet food, then they speak falsely, as they swarm in and out together. These, then I give you; enquire of them strictly and delight your heart: and if you should teach any mortal so to do often he will hear your response—if he have good fortune. Take these, Son of Maia, and tend the wild roving, horned oxen and patient mules.'*

(ll. 568a–573) *So he spake. And from heaven father Zeus himself gave confirmation to his words, and commanded that glorious Hermes should be lord over all birds of omen and grim-eyed lions, and boars with gleaming tusks, and over dogs and all flocks that the wide earth nourishes, and over all sheep; also that he only should be the appointed messenger to Hades, who, though he takes no gift, shall give him no mean prize.*

(ll. 574–578) *Thus the lord Apollo showed his kindness for the Son of Maia by all manner of friendship: and the Son of Cronos gave him grace besides. He consorts with all mortals and immortals: a little he profits, but continually throughout the dark night he cozens the tribes of mortal men.*

The hymn says that Hermes was still an infant when his brother the solar god gave him the bee maidens. They were to teach Hermes the art of divination and intuitive wisdom: of reading omens and signs (for example, in nature regarding the appearance and omens of the birds). This adds much more of a flavour of lunar consciousness (one that is diffuse and honouring the wisdom that comes in the night from the unconscious). Hermes was also designated the psychopomp or messenger of Hades who could journey to and from the underworld. He can engage with all mortals as well as all the gods: he can move in all realms. On the other hand, the art of Apollo was that of the soothsayer—knowing the wisdom of the all-powerful Olympian sky god Zeus. According to the hymn, it is honey that nourishes and inspires truth in the bee maidens and which makes the difference in the quality of the encounter with the gifts and qualities of Mercurius.

Honey, like Mercurius as *aqua permanens,* is the transformative substance. Without this ritual food of the gods being administered, the bee maidens' words and work lost its way, becoming disoriented and no longer truthful or authentic. However, the last line of the previous quote tells us that Hermes 'cozens' (that is, cheats or tricks) mortals in the dark of night. So, as in the psychological individuation process, the mortal individual needs ego strength and consciousness to orient oneself in this process, just as bees navigate by the angle of the sun and stay oriented to the sun and to their hive on their travels. Mercurius is the agent of change, 'profiting little', the hymn suggests, but ever challenging through the shadow and the unconscious.

Solar conscious and healing hit the mark like an arrow hitting the bulls eye, and thus they have come to indicate logos, solar clarity—as in direct or linear knowing and intervention, like that of allopathic medicine. On the other hand, the way of the bee maidens was linked to nature and the underworld, to the Great Mother and to more visceral, yet intuitive, feminine knowing—non-linear and, at best, the kind of lunar consciousness, fertilised by Mercurius, where solar and lunar are unified as Jung describes and evokes in *Mysterium Coniunctionis.*

Telesphoros, the central focal point of Jung's carved stone at his Bollingen tower, is the small hooded companion and helper of the god of healing, Asclepius (son of Apollo, the solar god). Telesphoros stands within a circle and holds a lantern. From this circle emanate the lines symbolising the four rivers that flow out of Paradise. The Latin inscription includes the words (translated here): 'This is Telesphoros, who roams through the dark regions of the cosmos and glows like a star out of the depths. He points the way to the gates of the sun and to the land of dreams'.[129] This kabir (a dwarf-like helper) is linked with the phallus or herm, and with the transformative and regenerative healing power of Mercurius. At the centre of the cloak of Telesphoros, Jung has carved the astrological glyph for Mercury.

Thus Mercurius is at the source—the wellspring—of healing. This is the kind of healing that honours dreams and holds the lantern of consciousness to help us understand the messages from the unconscious—from dreams, emerging from our incubation time (which is what that sleeping, 'perchance to dream', in the sacred temenos was called). The way of Mercurius as the *aqua permanens* in healing is in essence shamanic and thus deeply transformative. In myth, the bee is endowed with the qualities of the psychopomp, healer, and saviour of the cosmos too—wise, but ever in the flux of life and birth, death and initiatory rites of passage over thresholds. The way of Mercurius is potentially dangerous but also regenerative—he continually 'cozens the tribes of mortal men'.

---

129 *Memories, Dreams, Reflections,* p. 254. (In the 1989 revised Random House edition, it is on p. 227)

The metal Mercury is poisonous and, as its common name 'quicksilver' suggests, moves fast, breaking up and changing shape constantly—flowing and forming spheres or little globes.

Given the watery, flowing dynamic of this kind of redemptive healing, it is significant that the power of forecasting and making rain (that redemptive, descending moisture from the gods) was widely attributed to bees. Aristaeus, according to Kerényi, was closely associated with the honey aspect of Zeus and regarded as the Lord of Honey; he [Aristaeus] was initiated into the rites of Dionysos, and also performed rainmaking rituals. The rain is one form of communication from the gods. Symbolically bees are messengers of the gods and the rain is the message they bring to humankind and the earth, to moisten and to fertilise. Aristotle, Pliny, Varro, Virgil and others say the bees were weather prophets.[130] Gruppe, in relation to Greek mythology, also writes of this:

> *Droughts especially appear to have accompanied the spirits of the dead in bee-form, and for this reason the honey offering was almost always customary in rain-magic, and the power of predicting rain was attributed to the bee.*[131]

The life and life force of the bee and the hive are recognised as miraculous, industrious, dedicated and energetic, as well as transformative and regenerative. Bees are a kind of pontifix linking heaven and earth, and also the underworld. They, like Mercurius, are also described as having the role of the psychopomp—conductor of souls from this world to the next, venturing to the underworld, the realm of the dead. Bees, in some myths, can intercede for God and the life force by venturing to communicate with the devil and by journeying into the realm of the dead or Hell. Thus they symbolically link the highest heavens and lowest chthonic realm, evoking the connecting symbolism of the bridge and ladder, ascent and descent, like the angels on Jacob's ladder.

---

130 Aristotle, *Historia Animalium*, 627b; Pliny, *Natural History*, XI. 20; Varro. *On Farming*, III. 16, 37; Virgil, *The Georgics*, IV, 191. See also Ransome, *The Sacred Bee,* p. 109.

131 O. Gruppe, *Griechische Mythologie*, p. 801.

Creation Myths and Folk Stories

The making of the philosopher's stone or *lapis* in alchemy has long been viewed as a repetition of the Creation as well as the making of a glorified or 'resurrection' body (for example, the diamond or indestructible body, or immortal nucleus of the personality—in ancient Egypt and in Eastern religions like Buddhism).[132] Bees, the world's great pollinators, and thus a source of much nourishment, are often cast as the protagonists in creation myths. For example, in the following folk tale of the Kalahari Bushmen from southern Africa, the bee, with its qualities of working very hard to ensure that life and the world continue through its role in the creative seeding in the heart of a numinous, open flower, is of highest importance.

> *A mantis was carried by a bee over the darkened waters of the flood at the beginning of things. Worn-out and desperate for a place to rest with its heavy load in that formless waste, the bee saw a great white flower, half open above the flood. With its dying energy it laid Mantis to rest in the heart of the flower and placed the seed of the first spirit inside it, safe from the turbulent waters sweeping by outside.*[133]

In the churning flux the centre of the flower is reached at the still but fertile point of creative chaos. In their role as messengers between the worlds, the fire and energy of bees reflect something of solar fire or a divine golden spark. The bees thus seem to be a central symbol in the creation, salvation and understanding of the cosmos—a sacred centre around which life circulates and rotates.

The bond or relationship between the fertile life-giving and -carrying feminine and the growing, flowering and fruiting of plants is a beautiful image that is present at all stages of the development of symbolism. The psyche as flower (lotus, lily or rose; the virgin as flower in Eleusisa) symbolises the flowerlike unfolding or blossoming of the highest psychic and spiritual development. Birth from a blossom (Figure 14) is an archetypal symbol of divine origin—for example, of the solar divinities Re or

---

[132] See von Franz, *Creation Myths*, p. 222.

[133] Van der Post, Laurens. *The Heart of the Hunter*, p. 148.

Nefertem in Egypt); of the Buddhist 'divine treasure in the lotus';[134] or, in China and the modern West, of the Self in the Golden Flower.

The bee in mythology and folklore (particularly in creation myths) also mediates and moves as a psychopomp between the realms of not only the gods and the departed, but also of the devil as illustrated in the following story from Romania. Hermes/Mercury is the best known of psychopomps in Greco-Roman mythology. The story also tells us important details about the Mercurial nature of the bee, which bear witness to both the black and golden sides of the bee and its symbolism.

Figure 14. The Egyptian sun god's birth from a lotus flower.

134 C. G. Jung and Richard Wilhelm, *The Secret of the Golden Flower*.

Why is the Bee Black, and why is it making Honey?

How did the Bee outwit the Devil?

In the beginning only water and God and the devil existed. These two were all the time moving about upon the face of the waters.[135]

The story continues. God asks the devil to go to the bottom of the seas to bring up in God's name a handful of the seed of the earth. The devil does this but in his own name, returning to God empty handed as the sea had washed the seed away.

This happens three times, and on the third attempt the devil claims the seed in God's name as well as the devil's, not being willing to claim it in God's name alone. The creation myth continues with the devil trying to outwit God, and then the Earth expanding from these seeds to become exceedingly big.

Now of all the beasts and creatures which God had made, none were more pleasing in His sight than the bee, which was then playing in Paradise. The bee was white and not black as she is now, and I will tell you how she changed her colour.

God sent the bee to ask the devil what He was to do and what good advice he could give Him. The bee went at once as she was commanded, and came to the place where the devil lived. "Good morning, uncle," said the bee. "Good morning, sister," said the devil. "What has brought you to me?" "Well you see, God has sent me to ask you what He is to do with this huge Earth." But the devil grumpily and sneeringly replied, "If He is God, He ought to know better than ask a poor devil for advice. I am not going to tell Him; let Him find out for Himself."

The bee who was a clever little thing—it was not for nothing that God's choice had fallen upon her—pretended to fly away; but she soon crept back stealthily and settled noiselessly upon the beams of the door. She knew that the devil could not keep any secret, and that he would surely speak out. So, indeed it happened. No sooner did he think himself alone than he started chuckling and muttering to himself.

---

135 Moses Gaster, *Rumanian Bird and Beast Stories,* pp. 60f. There are many variants of this story.

"A clever man that God really is. He asks me what to do. Why does He not think of mountains and valleys?" (You must know that when the Earth was first made it was flat like a pancake.) "Let Him take the Earth in His arms and squeeze it a bit, then it will soon fit all right."

The bee overheard what the devil said, and rejoicing that she had got the answer, spread out her wings and started flying away. The buzzing of her wings betrayed her, and the devil, hearing the noise, rushed out of his cave with his whip in his hand, and said, "O, thou thief! So that is the way thou has cheated me. Mayst thou feed on what comes out of thy body!" And he struck at her with his whip. This changed the bee's body black: moreover, he hit her so badly that he nearly cut her in two. That is why the bee has such a narrow waist, so that she looks as if she were cut in two and barely hanging together by a thread.

Limping and sore, the bee came back to God, and told Him what she had overheard; God was greatly pleased, and, squeezing the Earth in His arms, He made hills and valleys and the Earth grew smaller. Then turning to the bee He said, "Out of thy body henceforth shall come only honey to sweeten the life of man, and he shall bless thee for thy gift; also thou shalt bring forth wax for candles for the altar." Then god went on to ask the bee what reward she would claim for the errand she had so well fulfilled. The bee, impudent and greedy, replied, "Why should man share my gift and have my honey? Give me the power to kill with my sting."

And God was angry with the impudence of the bee, and replied: "All the honey shall be thine alone if thou art able to make a gallon of it in the summer; if not, man may share it with thee. And because thou hast asked for the power of killing with thy sting, meaning to kill men with it, thine own death shall be in thy sting."

This is the reason why the bee works so industriously and indefatigably during the summer. Each hopes to make a gallon of honey and never succeeds, and this is the reason why the bee dies if it stings anyone.

****

The bee has a very powerful creative energy but, as the above story shows, the bee is not all 'sweetness and light'. It has a tricky, unpredictably fierce, angry and driven aspect. It is indeed Mercurial. The bee's sting is indeed a force to be reckoned with. Mostly, when a bee stings it dies. The fertilised queen, whose sting is barbless, can sting many times, but usually does so only to stop another queen developing or to kill drones when they have no further purpose within a particular hive. The worker bees only die after stinging a threatening human or fleshy mammal; as when the bee tries to fly away, her sting remains embedded. Then she dies within seconds, as this dismemberment is too great to survive. If she were to sting another insect, for example, she would not die because the embedding and tearing would not happen. It is the piercing of the flesh, the getting under the skin, the intensity, that causes death. The story portrays the bee as having a duplex Lucifer/Satan flavour, not unlike that of Mercurius.

In terms of creation, consider that bees are often linked with lions and bulls in mythology (Figure 15). The lion is the animal of the sun in astrology (as the sun is the planetary ruler of Leo, the fixed fire sign) and the bull is the animal of Venus (as ruler of Taurus, fixed earth). The fixed nature of this energy means that something solid, objective or concrete is constellated or is capable of being created.

Two powerful and poetic descriptions of bee stings illustrate the strength of the impact of this little insect on humans if they are stung. Maurice Maeterlinck writes:

> *Her [the worker bee's] sting, which produces a pain so characteristic that one knows not where with to compare it; a kind of destroying dryness, a flame of the desert, rushing over the wounded limbs, as though these daughters of the sun had distilled a dazzling poison from their father's angry rays.*[136]

Ted Hughes describes in his poem 'The Bee God' how bees 'planted their volts' in him, their target, 'seering, jolting and electrifying'.[137] This impact has often been compared with that of the arrows of Cupid or Eros piercing the heart of the love stricken. So we are helped to viscerally understand that the bees can represent a fierce and burning dynamic. Their power is indeed great, like that of Mercurius.

---

136 *The Life of the Bee,* p. 25.

137 See 'The Bee God' in *Birthday Letters,* p. 151.

Figure 15. 'Venus, the Subject of the Wise, displays her wondrous qualities. Everything proceeds from her.' Michael Maier's illustration in *Tripus aureus* (The Golden Tripod), Basilio Valentino, Frankfurt, 1618. (The bee is represented by Cupid.)

3
Bee Symbolism and the Religious

Thou seemest—a little deity!
—Anacreon, Ode 34: 'To the bee' (fifth century B.C.).

And he [Jonathan] put forth the end of the rod that was in his hand,
and dipped it in the honeycomb and put his hand to his mouth
and his eyes were enlightened.
—1 Samuel 14: 27.

The first time we open a hive there comes over us an emotion
akin to that we might feel at profaning some unknown object,
charged perhaps with dreadful surprise.
—Maurice Maeterlinck.

Many attributes of bees, honey and the hive have long been regarded as sacred and numinous. I have already mentioned various sacred contexts, and it is the numinous that forms the warp and weft of this written tapestry. This section will focus on some of the more colourful and vital examples of bearing witness to the central significance of the bee, the honey and the hive in the context of the major world religions, but will not exclude the more chthonic and pantheist religions of earlier times. Many poets and writers, including prophets or mystics, have waxed lyrical about the bee. Virgil, in his great epic of the bees, paid tribute to their divine nature and resilience, saying (or singing):

(Some) have declared that on bees is bestowed some share in the soul divine,
Some draughts of the stars of heaven; for the god moves everywhere
Through earth, the expanses of sea and the limitless depth of the air.
Yea, and to Him they return, for not unto Him do they die
At dissolution! There is no death; but they live and they fly
To the ranks of starland and enter the high-reared hall of the sky![138]

---

[138] Virgil, in A. S. Way, trans., *The Georgics,* IV, 220 ff.

The bees have a long tradition of being dear to the gods and numinous to humankind. Through their honey, they bring us sweetness but also teach that a powerful (in some cases, deadly) sting is needed to protect the honey and the triumphant and indestructible life force represented by the queen and the self-regulation of the hive. Bees have long been regarded as bringing this same sweetness to the gods. Nectar, 'the food of the gods' from the centre of flowers, is the substance that is sought by the bees. Humans bring offerings of sweetness to the gods, as in Eastern religions like Buddhism. We also offer our intimate loved ones sweet foods, perfumes and flowers.

In the context of world religions, the bee and honey since ancient times have been accorded great significance. This little honey-making insect, without tiring, works the centre of the flowers all day long and then retires into the centre of the hive in the hours of darkness before starting out again next day to begin the next journey to the centre of the blossoms. All this is done to serve the queen bee, the royal centre, who through her fecundity enshrines the life force and vitality of the hive and thus the indestructibility of life: the *zöe.* The natural rhythms of the life force of the hive and, by amplification, the symbolic nature of the bee, the honey and the hive enact the circumambulation of a sacred centre.

Honey and the Religious

Honey as a fundamental or prime source of nourishment often goes together with milk as the food which welcomes infants, mortal and immortal, into the world. It also helps to initiate the spiritually reborn across thresholds. These two foods are sacred in that they are unalloyed: pure, natural, nourishing and supremely supportive of the life force.

Our communal life is more complicated than that of bees, but we also have an instinctual basis for it. Our instinctual basis was, and still is, reflected in the performing of rituals that hark back to the most archaic times.[139] In the rites of passage associated with birth, marriage and death, the gods are often offered honey and, in reverse, the divine powers of the gods were often called upon to give moisture, for example as rain

---

139 Von Franz, *Shadow and Evil in Fairy Tales,* pp. 52f.

or distilled honey from the clouds to help crops like corn to sprout.[140] The bees in a particular hive show a remarkable adaptability to the new—they do not stagnate. There is great fertility and renewal, as in the symbolic regeneration in the sprouting of grain from the corpse in Pharaonic Egypt.

The perfection and miraculous nature of honey, the sweet and mysterious golden healing substance, make it a central component of many religious rituals, for instance as an offering to supreme deities and fertility spirits. I will describe briefly some specific examples in this section, but the religious is a thread that runs throughout the discussion of the vast symbolism of bees, honey and the hive.

****

Honey is the one food that never putrefies. It can be eaten after centuries. In ancient times it was also used as an effective embalming fluid, for it is excellent at killing bacteria. Bee Wilson expresses the fact well, saying that 'in former ages honey was always a way of joining the living with the dead'.[141] Burial in honey dates back at least 4,000 years, and honey itself was a very significant element in ancient death rituals. It had several roles including preserving the body, 'exorcising evil spirits and feeding good spirits, nourishing the mourners, sweetening the mood of the gods, and adorning the consecratory rites. In an ancient funeral hymn of the *Rig-Veda,* priests are directed to pour out honey for the god of death. In India, honey was a good funeral offering because it was a 'life substance' and to offer it to the gods was therefore fitting. "It is bee's honey they say; for bee's honey meant the sacrifice," intone the Satapatha Brahmana.'[142] Also in many cultures honey was considered to be a valuable funeral food because it symbolised both life and death.

Honey as a sacred substance is also used in marriage rituals. There are many more examples of honey or nectar as a sacred ritual food, often as honey cakes or mead. I will include but a few to illustrate their diversity and some aspects that relate specifically to certain religions.

---

140 Ransome, *The Sacred Bee,* p. 185.

141 Wilson, *The Hive,* p. 190.

142 A. E. Fife, 'The concept of the sacredness of bees, honey and wax', p. 42.

The legendary parthenogenic origin of bees has also contributed to a view of honey as a miraculous, uncontaminated sacred food.[143] In her commentary on the visions of Saint Perpetua, Marie-Louise von Franz refers to the words from a magic papyrus: 'Drink milk and honey before sunrise, and in thy heart there will be something divine', and reminds us that in ancient Egyptian initiation rituals neophytes were first given a cup of water, then a mixture of honey and milk, and that finally wine and water became the real Eucharist.[144]

We have already seen the importance of the bee in the cult of the Great Mother. In ancient Crete, she took the form of the Bee Goddess. Zeus and Dionysos were fed honey from birth. The role of honey as a way opener to the wisdom of the unconscious (for example, when consulting the oracle at Delphi or in an incubation rite like those of Trophonius or at the sacred Asclepia evoking the healing through dreams has previously been described here. In the Eleusinian Mysteries honey was given to the higher grade of initiate as a sign of new life. For the Minoans, honey had an important role in ritual as a food for both the living and the dead. Honey was offered to Mithras, and this was probably linked to the bees as being 'the stars of heaven'. Honey was poured onto the hands and tongue of the initiated.

For some Amerindians, the bee and its food (pollen and honey) have a role in the pollen path ceremonies and in the rituals of initiation and purification. Mead, a sacred fermented honey drink, was particularly important in the ceremonies of the Celtic druids.

Honey is of central significance in the marriage customs in many lands including India, Greece and Turkey. Rituals and the customs related to weddings and the honeymoon involved honey. The doorpost and lintels of the newlyweds' home would be smeared with honey and also the bride's lips or genitals. These rituals were often presided over by the bridegroom's mother. Among Albanian gypsies and at weddings in some parts of Turkey the bride was given honey and butter with which she smeared the doorposts of her new home. This custom is probably a very

---

143 J. C. Cooper, *An Illustrated Encyclopaedia of Traditional Symbols,* p. 84.

144 Von Franz, *The Passion of Perpetua,* p. 45.

ancient one and is still found in modern Greece and in other countries where the wedding guests throw sweets at a newly married couple. This association of sweetness with the ritual blessing of a new marriage had its roots in antiquity.

Another variation was that when the bridal pair first arrived at the bridegroom's cottage on the wedding day, the groom's mother would be waiting there holding a glass of honey and water. The bride then had to drink from this glass so that the words on her lips might become as sweet as honey. The lintel of the door was then daubed with the remainder of the honey, so that strife might never enter this particular house.[145]

Such customs also recognised that too much honey (or honeymoon) led to the dark side of the sweetness—the laziness and matter-bound inertia or self-indulgence as described in the Grimm's fairy tale 'Lazy Harry'. Honey in its negative aspect, as commented on by Paracelsus, 'becomes a deadly poison, like *Tartarus mortalis'*.[146] Too much honey or sticky sweetness can be cloying and dangerously seductive, like 'the honey that pours from a courtesan's lips' (Proverbs 5: 3). But too much honey is what you see in the love stories of Samson and Delilah and of Venus and Adonis. Adonis is a beautiful youth content to vegetate in the arms of his lover. He is short lived like the so-called 'gardens of Adonis' whose plants, the lettuce and anemone, cannot withstand the elements for long.[147] Their opposite, the gardens of Demeter, endure, as they are the product of more differentiated (or, in depth psychological terms, individuated) experience. The honey in the hive is not free flowing; it is held in the hexagonal comb and capped with wax. Bees really value it and, as we have seen, defend it fiercely.

---

145 Ransome, *The Sacred Bee,* p. 187; from Rodd Rennell, *Customs and Lore of Modern Greece.*

146 *Mysterium Coniunctionis,* CW 14, par. 682.

147 See M. Detienne, *The Gardens of Adonis*: *Spices in Greek Mythology;*. See also von Franz, *Shadow and Evil in Fairy Tales*, p. 44, regarding the Attis-type, the beloved son of the Great Mother who does not grow old and decay. This type represents the pattern of the *puer aeternus*, the perennially young god, about whom von Franz comments: 'Like this god, most young men who have a decided mother complex feel, at some moment in their lives, that the process of life does not allow for such a state of being eternal; it has to die.'

****

The ancient Greeks often embalmed the dead by preserving them in honey. The bodies were arranged crouched in large vases in the foetal position ready for their next birth. Sybille Birkhäuser-Oeri wrote of how close the beginning of the life cycle or individuation process is to the end:

> *The maternal principle, although or even because it symbolises the unconscious, is also what gives birth, creates existence. So there are in fact two essential sides of the maternal: one which is trying to create consciousness and existence and uses all available means to achieve that goal, and another which seeks unconsciousness and non-existence, in other words, destruction and disintegration.*[148]

Also in ancient Greece, where honey was thought to rain down from the heavens, bees were highly regarded as the collectors and distributors of the nectar, and in this role were important mediators and messengers between those who lived on earth and the sky gods. It was important to be on good terms with the bees if you needed anything from these deities.[149]

Ancient Egypt: Re's Tears and 'Opening of the Mouth' Ceremony

In the extensive Nile Delta, bee cultivation or apiculture has been in evidence since the third millennium B.C.,[150] and honey has long been used both as a food and in medicine (for example, as ointments). Many traditions and customs relating to beekeeping have remained more or less constant since beekeeping began in Egypt 4000 to 4500 years ago. Beekeeper and bee historian Eva Crane mentions that she was sent details of a papyrus (Cat. I 384) from 404–343 B.C. known as 'The Myth of the Eye of Re', consisting of popular stories (somewhat like Aesop's fables). According to this papyrus, the beekeeper called the bees with a reed flute when he wanted them to swarm. This 'is because a reed flute excited Nut

---

148 Birkhäuser-Oeri, *The Mother: Archetypal Image in Fairy Tales,* p. 22.

149 Bishop, H. *Robbing the Bees: A Biography of Honey the Sweet Liquid Gold that Seduced the World,* p. 21.

150 Crane, *A Book of Honey,* p. 251.

[the sky-goddess, see Figure 16]. In Ancient times . . . If one wants to write the word 'honey', one draws Nut holding a reed in her hand'.[151]

In ancient Egypt, Re was the sun god (born, according to one of the official legends, from Nut every morning) and when he wept his tears turned into golden bees (as golden dew, rain or sun rays) raining down on earth in a redemptive way. Representations of bees can be found at Re's temples.

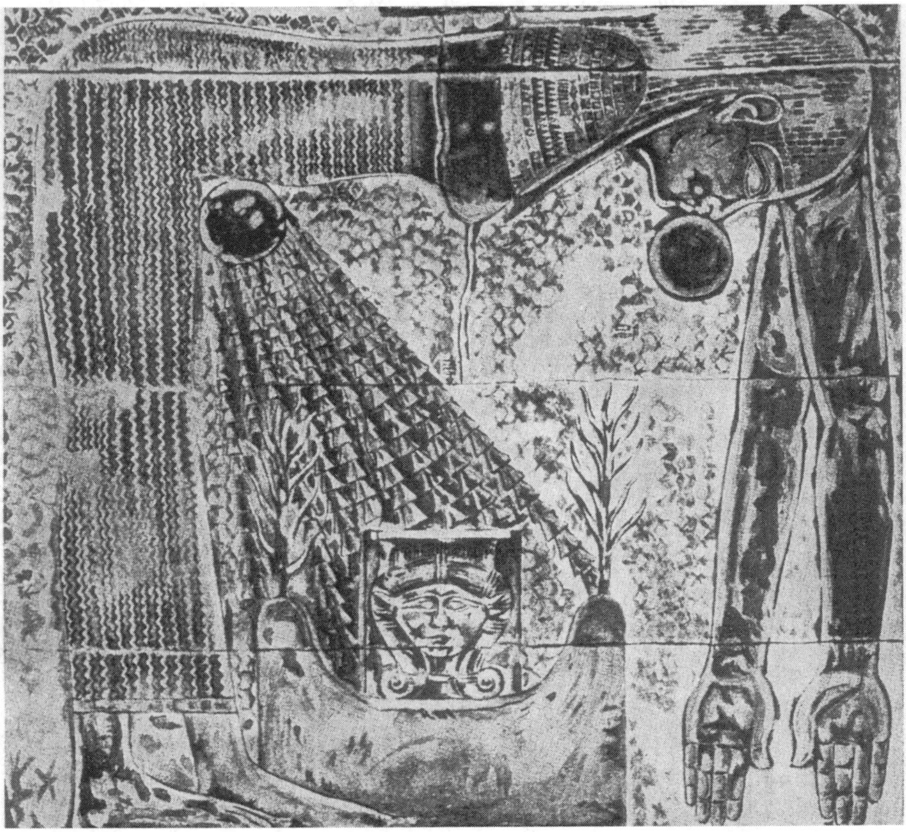

Figure 16. The Egyptian Goddess Nut, painted ceiling relief; Temple of Hathor, Dendera, Egypt.

151 Ibid, p. 170.

A papyrus in the British Museum reads:

> The god Re wept and the tears
> From his eyes fell on the ground
> And turned into a bee
> The bee made [its comb]
> And busied himself
> With the flowers of every plant;
> And so wax was made
> And also honey
> Out of the tears of the god Re.[152]

This verse recounts a creation story or myth of the bees, who after being created from the eye of the sun god then became creators themselves. Honey and beeswax thus had the same sacred origin in the tears of the god Re.

As Hilda Ransome writes in her extensive work *The Sacred Bee in Ancient Times and Folklore,* almost all the religious rites of the Egyptians were concerned with the afterlife. After the process of mummification, there was an elaborate ceremony called the 'Opening of the Mouth', in which honey was an important offering. The imagery of the following contemporary archetypal dream resonates with this vivifying ancient rite.

> *I was sitting in nature on a single rock. [This rock reminded the dreamer of the rock Jung writes about from his childhood in* Memories, Dreams, Reflections.] *Then it became some kind of initiation in a dark enclosed cave-like place with rock walls. I was lying on my back on a bed or couch or even a rock ledge in an alcove next to a wall made of stone or rock. Priests (ancient Egyptian ones) opened my mouth with a spatula-like ritual instrument and fed me honey, which they placed on my tongue.*

In the Egyptian ceremony, priests used special spatula-like instruments to 'open the mouth' of the statue of a god, and of the statue or mummy of the king or other deceased noble. In one part of the ritual, the Kher-heb, the officiating priest, says the words, 'Going about as a bee, thou seest all the goings about of my father [Re the sun god].'[153]

---

152 Ransome, *The Sacred Bee,* p. 33.

153 Ibid., p. 31.

Given the ancient belief found in many parts of the world that souls take the form of bees, Ransome says:

It is just possible that this text may contain the germ of this belief. The Egyptians believed that a man possessed a ka or double, who had to be fed after death and this going about like a bee may refer to this ka. It is also to be remembered that the symbol for the Pharaoh as the king of Lower Egypt was the bee, so it would be natural for the people to think of his ka, or that of one of his subjects, appearing as such.[154]

The following text was also included in the Opening of the Mouth ceremony, which was performed daily in the Temple of Amun-Re. One part of the service was called the 'Chapter of the Festal Perfume in the Form of Honey', and in it the Kher-heb recited the following passage:

Hail, Amen-Ra, Lord of the Throne of the Two Lands! I present unto thee honey, the Eye of Horus, the Sweet One, the exudation from the Eye of Ra, the Lord of Offerings, Amen-Ra, the Lord of the Two Lands is flooded therewith, for it is sweet to thine heart, and it shall never depart from thee.[155]

Another passage related to this is found in the Salt Magical Papyrus:

When Ra weeps again the water which flows from his eye upon the ground turns into working bees; they work in flowers and trees of every kind and wax and honey come into being.[156]

Seen psychologically, then, the sun is a symbol of the source of awareness and guarantees all order. The longing for light is the longing for consciousness and the new sunrise was a resurrection, when the rejuvenated sun god returns to life.[157] For the ancient Egyptians, participation in the life of the sun meant, therefore, participation also in the increasing process of becoming conscious and of the evolving culture of humankind.[158]

---

154 Ibid, pp. 31f.

155 From Wallis Budge, *Opening the Mouth*, vol. ii, p. 206.

156 British Museum, No. 10051, recto 2, 5–6.

157 Von Franz, M-L. *On Dreams and Death*, pp. 66–7.

158 Alchemy and the individuation process of Jungian or depth psychology are strongly

Judaism: Wisdom, Sweetness, Samson and Lion's Honey

The great natural wisdom of the bee is recognised in the following moving tale:

> *It is said by the Rabbis that when the Queen of Sheba visited King Solomon she held out to him two wreaths, one of artificial flowers and the other of genuine flowers. Unable to discover which were the real flowers, the king caused the casement window to be opened, when, lo! A flight of bees entered and lighted on the natural flowers!*
>
> *The bees were wiser than the wise king!*[159]

The symbolism of honey is used to illustrate ethical teachings in various religious traditions. The Bible includes the following: humans are exhorted to eat honey and the honeycomb (Proverbs 24:13), but warned against overdoing this (Proverbs 25:16-27). Honey is a simile for moral sweetness (Ezekiel 3:3), and for the excellence of the law (Psalm 19:10), for pleasant words (Proverbs 16:24), and for love and sweet words.

In the Old Testament of the Bible, the Book of Judges 13-16 tells the story of Samson who was born after an angel of the Lord visited his barren mother and prophesied his birth. In Judges 14:5-9, the story reaches one of its symbolic climaxes:

> Then went Samson down, and his father and his mother to Timnath, and came to the vineyards of Timnath: and, behold, a young lion roared against him.
>
> And the Spirit of the Lord came mightily upon him, and he rent him [the lion] as he would have rent a kid, and he had nothing in his hand: but he told not his father or his mother what he had done.
>
> And he went down, and talked with the woman; and she pleased Samson well.
>
> And after a time he returned to take her, and he turned aside to see the

---

connected symbolically with the ancient Egyptian religious transformation processes of death and regeneration symbolised by the daily journey of the sun-god dying into darkness and his return as Kephri, the new-born sun. For further reading see Theodor Abt and Eric Hornung, *Knowledge for the Afterlife: The Egyptian* Amduat—*a quest for immortality*, and Andreas Schweizer, *The Sungod's Journey through the Netherworld: Reading the Ancient Egyptian* Amduat.

159 Ransome, *The Sacred Bee,* p. 74.

carcass of the lion: and behold, there was a swarm of bees and honey in the carcass of the lion.

And he took thereof in his hands, and he went on eating, and came to his father and mother, and he gave them, and they did eat: but he told not them that he had taken the honey out of the carcass of the lion.

Samson, whose name in Hebrew means little sun, was inspired by this experience and created the following riddle that then sealed his fate as an exceptionally strong man, physically speaking: 'Out of the eater came forth meat, and out of the strong came forth sweetness.' (Judges 14:14).

In the story of Samson, the strength of the lion becomes bees in their dripping honeycomb. This is linked with the *bugonia* (the ox- or bull-born bees). The carcasses of big powerful animals turn into a multiplicity of bees and golden honey, which is the food of the bees and of the gods. Edward Edinger says of this, that 'the changing of the lion into honey is a symbolic image of the transformation mystery—the transformation of power into sweetness.'[160] The riddle Samson created and the marvellous honey he tasted really struck him with awe. It was numinous to him, as was the angel announcing to his mother that Samson would be born. The bees and honey brought Samson new enlightenment, inspiration and eros, as well as the potential for embodied healing and sweetness. The mystery of the honey was a way of bridging the opposites of pain and bitterness that Samson experienced, and of transforming them.

He got stuck however and went to sleep in Delilah's lap, only to have the god-given strength of his wild Nazarene locks of hair removed as he slept. He did not become a warrior sage uniting heaven and earth.

Christianity: Bees, the Hive, the Honey and the Holy Mother

Bees depicted on early Christian tombs were emblematic of everlasting life, presumably because they symbolised virtue rewarded and because during three months of the year (winter) they disappeared from life, as Christ did for three days after the crucifixion and before the resurrection. The reappearance of the bees in spring is thus equated with the Resurrection.[161]

---

160 Edinger, *The Mysterium Lectures*, p. 274.

161 Stevens, *Ariadne's Clue*, p. 365.

Figure 17. Images painted on a Carniolan hive front from 1869, showing Mary with the crowned Christ-child.

After swarming, bees start to lay down the honeycomb's fundamental structure as soon as the swarm lands. The cluster of bees is like the grapes of Christian, pagan and alchemical iconography. The candles used in Christian churches has long been made with bees' wax. In early times this wax, regarded as 'being produced in such fertile profusion by chaste bees, came to symbolise the most significant chastity of all: that of Mary mother of God'. Like Mary, the wax-producing bees possess a mysterious combination of fecundity and virginity. Bees 'produce posterity, rejoice in offspring, yet retain their virginity', says the hymn that is sung on Holy Saturday (the day before Easter) honouring the combination of those extreme alchemical opposites fire and water, in the Paschal candle.[162] This is part of a special ceremony, the Benedictio Fontis, in which the baptismal water in the font is consecrated and transformed into the 'regenerating and purifying fountain of life, the immaculate womb of the divine source' and thus can regenerate human individuals as spiritual beings.[163]

---

162 Wilson, *The Hive,* p. 83, and Eva Crane, *A Book of Honey*, p. 138.

163 'The Tavistock Lectures', *The Symbolic Life,* CW 18, par. 364.

Figure 18. The third of six scenes from a Galician manuscript, ‘Hymns to the Virgin’, late 1200s or earlier, in the Library of El Escorial, Spain. The bee-keeper, and perhaps his wife, discovering the Virgin and Child in a hive in which he has hidden the consecrated Host.

Figure 19. 'I turn everything to honey'; image from the church roof, Wallfahrtskirche, Hergiswald, Switzerland.

In Christianity, honey is also linked with the earthly ministry of Christ and the divine word. Bees, and particularly the queen and the community of the hive, are associated with the Virgin Mary (whose image with the Christ child was painted on wooden beehives in Central Europe, for example). Mary is also symbolically a closed garden, like a contained sacred walled garden with a fountain in the centre. Figure 19 (page 99) shows a beehive image from the ceiling of the church at Hergiswald in Switzerland. It bears the following divine message in Latin: 'I turn everything to honey'.[164] Mary's was the corporeal body that gave birth to the Christ child, the incarnation of the deity.[165]

In alchemical terms, the Assumption of the Virgin Mary symbolises the 'white foliated earth', that is, the entry of the purified principle of materiality into the conception of totality or the assimilation of the principle of materiality (of the body, of incarnated existence) into the Western psyche. Psychologically this is like the stage reached after a thorough analysis when matters related to the personal unconscious and the shadow have been purified and are ready to be incorporated into the conception of totality.[166]

The sixteenth-century Christian mystic imagery of St John of the Cross's Spiritual Canticle 27:8 also wonderfully evokes the transforming eros nature of the bee:

> *Like the bee that sucks honey from all the wild flowers and will not use them for anything else, the soul easily extracts the sweetness of love from all the things that happen to her, that is, she loves God in them. Thus everything leads her to love.*[167]

Acclaimed Russian writer Leo Tolstoy is famed for lyrically expressing the rhythms and numinosity of living a simple life close to nature. Tolstoy takes up the metaphor of the questing, swarming bees to paint a persuasive image of how the collective stasis could be broken and transformed: one cohesive cluster breaks up, searches for a new resting place

---

164 *Encyclopaedia of Traditional Symbols,* p. 84.

165 Edinger, *The Mysterium Lectures*, p. 132.

166 Ibid.

167 St. John of the Cross, Spiritual Canticle 27:8.

and focus; the bees leave the old hive in the search for a new home, before regrouping around a new centre.

In his piously titled 1893 work, 'The Kingdom of God is Within You: Christianity not as a Mystic Religion but as a New Way of Life', Tolstoy claimed that men and women could find a new meaning and centre of devotion in their daily lives. To him, the imperative of the swarm of bees (the questing cluster of bees flying loosely at first but quickly regrouping to centre protectively around the queen) to find a new place to settle and re-establish their community was just like humans needing to find a new place to live or find sanctuary—for him and many Christians, this was the place of god.[168] The bees could escape their temporary homelessness, Tolstoy said, because, each of them is 'a living, separate creature endowed with wings of its own' with which to fly to God and live the life of a Christian. Tolstoy wrote:

> *Only let one bee spread her wings, start off, and fly away, and after her another, and another, and the clinging, inert cluster would become a freely flying swarm of bees. Just in the same way, only let one man look at life as Christianity teaches him to look at it, and after him let another do the same, and the enchanted circle of existence in the state conception of life, from which there seemed no escape, will be broken through.*[169]

Bee Wilson aptly points out that in this text Tolstoy is expressing a belief that 'the bees have a message for us, if only we can grasp it—the message of Christianity. To which we might reply, 'But what about all the other messages of the hive? The message to work? To love? To live? To die? To govern? To eat? To dance?'[170]

I would add: 'To have a nourishing and life-renewing contact with a sacred centre daily', as this dynamic has a parallel in the psychological symbol of inner order—the mandala—and in the circumambulation of the Self—the life-renewing god-image within.

---

168 Wilson, *The Hive,* p. 270.

169 Tolstoy, 'The Kingdom of God is Within You', chap. 4.

170 Wilson, *The Hive,* p. 270.

Figure 20. *Dat Rosa Mel Apibus* (The rose gives honey to the bee); engraving from Robert Fludd, *Summum bonum,* Frankfurt, 1629.

Cathedral Builders: Bees and Freemasons

Bees are builders extraordinaire. Their golden honeycombed homes have often been compared to cathedrals—dimly lit with a golden glow. The wonder of the hexagonal precision defies understanding. It is a miracle and a mystery of nature. The freemasons, originally stonemasons who were the great cathedral builders, identified with the bees in various symbolic ways.

The Freemasons' Monitor lists a series of emblems important to masons. The most extensive commentary is on the beehive, which is intrinsically connected with 'the bee that makes and inhabits the hive, and the honey which is the product'. For example:

> *The Bee Hive is an emblem of industry, and recommends the practice of that virtue to all created beings, from the highest seraph in heaven to the lowest reptile in the dust. It teaches us that we come into the world rational and intelligent beings, so we should ever be industrious ones;*

never sitting down content while our fellow creatures around us are in want, especially when it is in our power to relieve them without inconvenience to ourselves.[171]

Figure 21. Alchemical image of bees and beehive; frontispiece from *Geheimnisse einger Philosophum und Adepten* (Secrets of Philosophers and Adepts), 1780.

171 A. E. Roberts, *The Craft and Its Symbols,* p. 74.

Other Monitors give similar explanations that emphasise the themes of industry and a strong work ethic, social harmony and cooperation. Allen Roberts comments on the beehive becoming an important part in Freemasonry as it:

> *symbolised what the cathedral builders did and the way they did it. The bee definitely is industrious. He works hard and tirelessly, not for himself, but for the swarm. He has a strength and knowledge of materials that cannot be duplicated. He works in complete cooperation, and without dissention, with his fellow bees. He protects the Queen, refuses admittance to enemies, builds, makes honey, and lives in a society ruled by law. What bees do can be compared with the cathedral builders of centuries ago. . . . Undoubtedly, the operative masons saw their duplicate in the bees.*[172]

Islam: 'The Bee' Sura

In monotheistic Islam there is no mythological or cult reverence for the bee, but its work and industry were highly esteemed. Writer Ibn Magih quotes Mohammed as saying: 'Honey is a remedy for every illness, and the Koran (Qu'ran) is a remedy for all illnesses of the mind, therefore I recommend to you both remedies, the Koran and honey'.[173]

The Koran says some important things about bees and Sura 16 is called 'The Bee'. Sura 16:69 includes the following lines:

> *Thy Lord has taught the bee saying, 'Provide thee houses in the mountains and in the trees, and in the hives which men do build for thee. Feed, moreover, on every kind of fruit, and walk the beaten paths of the Lord.' From its belly cometh forth a fluid of varying hues, which yieldeth medicine for men. Verily in this is a sign for those who consider.*[174]

Another indication of the bee's great significance is that the prophet Mohammed designates the bee as the only animal that is addressed by Allah himself. An old Arabic commentator on the above passage says that honey is of various colours according to the flowers the bees visit, adding that honey is not only good for eating, but is also highly useful

---

172 Ibid.

173 Hilda Ransome, *The Sacred Bee,* p. 73.

174 Ibid., p. 71.

medicinally and relates the following story:

> *A man went to Mohammed and told him his brother had violent pains in his body, and the Prophet told him to give the sick man honey. He did as he was told, but soon came back to say his brother was no better. Mohammed answered, "Go back and give him more honey, for God speaks the truth; thy brother's body lies." When the honey was taken again, the sick man, thanks to the grace of God, recovered immediately.*[175]

In almost all cultures the healing qualities of honey are venerated. In Islam, according to the Prophet, honey restores sight, preserves health and resuscitates the dead.[176]

Hinduism: The Blue Bee and Kama the God of Love

The Sanskrit word for honey is *madhu* and the gods Vishnu, Krishna and Indra were called *Madharva,* the nectar-born ones. Their symbol is the bee. Vishnu is sometimes represented as a blue bee resting on a lotus flower—an ancient symbol of life, of resurrection, and of Nature whose power slumbers until the warmth and light of the sun calls it to life, like the bee. Shiva was depicted in a less terrible form as *Madheri* ('The Suave One'), a bee on a triangle.

Kama, the Hindu god of love, after whom the *Kama Sutra* is named, is often depicted (for example on Indian honey jars) riding or flying holding onto a lion-like creature, which in turn is supported by a flying bee (Figure 22, next page).[177] Kama also uses chains of strings of bees as missiles shot from the 'bow of love'. The fifth-century Indian poet Kalisasa wrote:

> *A stalwart soldier comes, the spring,*
> *Who bears the bow of Love:*
> *And on the bow, the lustrous string*
> *Is made of bees.*[178]

---

175 Ibid., pp. 71f.

176 Ibid.

177 Ransome, *The Sacred Bee,* pp. 45f.

178 Ibid., p. 45.

Kama's bees, like Cupid's, symbolise that eros or love has both the honey sweetness and the passionate drive and sting. Both the Vedic moon god and the moon's elixir were called Soma. The nectar, like Soma, was 'the honey of all beings', and mortals would drink this 'honey' in ceremonies to become one with the eternal: 'I have drunk Soma. I have seen the gods'.[179]

Figure 22. The Indian god Kama supported by a bee.[180]

179 Thanks to Jules Cashford for lunar inspiration. See her book *The Moon: Myth and Image*.

180 Ransome, *The Sacred Bee,* p. 46.

Hindu sacred texts also include wonderful insights into how fundamental a substance honey is. In the *Rig-Veda,* honey is regarded as a fecundating principle, a source of life and immortality. In the *Brihadaranyaka Upanishad*, the Fifth Brahmana includes fifteen stanzas, each of which begins with 'this *earth*, ... (or this *water*, *fire*, *air*, *sun*, *space*, *moon*, *lightning*, *thunder*, *ether*, *law* (dharma), *truth* (satyam), *mankind*, *Self*), for example:

> *1. This earth is the honey (madhu, the effect) of all beings, and all beings are the honey (madhu, the effect) of this earth. Likewise this bright, immortal person in this earth, and that bright immortal person incorporated in the body (both are madhu). He indeed is the same as that Self, that Immortal, that Brahman, that All.*
>
> *2. This water is the honey of all beings, and all beings are the honey of this water. Likewise this bright, immortal person in this water, and that bright, immortal person, existing as seed in the body (both are madhu). He indeed is the same as that Self, that Immortal, that Brahman, that All.*
>
> *14. This Self is the honey of all beings, and all beings are the honey of this Self. Likewise this bright, immortal person in this Self, and that bright, immortal person, the Self (both are madhu). He indeed is the same as that Self, that Immortal, that Brahman, that All.*
>
> *15. And verily this Self is the lord of all beings, the king of all beings. And as all spokes are contained in the axle and in the rim of a wheel, all beings, and all those selfs [of the earth, water etc.] are contained in that Self.*[181]

Raymond van Over, as editor of the compilation of sacred texts *Eastern Mysticism*,[182] included the following footnote to the Fifth Brahmana to help further understanding of this sacred honey text:

> *Madhu, honey, seems to be taken here as an instance of something which is both cause and effect, or rather of things which are mutually dependent on each other, or cannot exist without one other. As the bees make the*

---

181 *Brihandaranyaka Upanishad,* Fifth Brahmana, stanzas 1-15.

182 His compilation contains texts from 'The Near East and India'. Heartfelt thanks to Anne Baring for leading me to this reference. Marie-Louise von Franz included a version of the stanza on the Self in her Transformed Berserker lecture, and her citing of this Brahmana was a key source of inspiration for me.

honey, and the honey makes or supports the bees, bees and honey are both cause and effect, or in all events are mutually dependent on one other. In the same way the earth and all living beings are looked upon as mutually dependent, living beings presupposing the earth, and the earth presupposing living beings. This in all events seems to be the general idea of what is called the Madhu-vidyâ, the science of honey, which Dadhyak communicated to the Asvins.[183]

The *Madhu-vidyâ,* the science of honey, which Dadhyak communicated to the Asvins, follows in these stanzas:

16. Verily Dadhyak Âtharvana proclaimed this honey (the Madhu-vidyâ) *to the two Asvins, and a Rishi, seeing this, said [from* Rig-Veda *I, 116, 12]:*

The Self is honey for all beings. For this Self all beings are Honey. And that which in this Self is that atman—that is, that purusha that arises out of light energy, out of the deathless—is that very primordial Atman Deathlessness, Brahman. It is the universe. And verily is the Self the lord of all beings, the king of all beings. And just as all the spokes are held in the axle and the rim of a wheel, so all beings and all these selves [of the earth, of the waters, etc.] are held in the Self.[184]

Marie-Louise von Franz included the above passage from the *Rig-Veda* as central to her lecture 'The Transformed Berserker', at the Eighth Conference of the International Transpersonal Association of Individual Transformation and Universal Responsibility (Davos, Switzerland, 1983. She emphasised that in India, *madhu* (honey) symbolises the contact of all beings in the universe with the Self, the anthropos *(purusha)*—an objective, complete and mutual interdependence or connectedness of all things.

This is what Jung calls 'objective knowledge' and 'the eternal connection through fate' as distinct from subjective love with its psychological projections and ego-oriented wishes.[185]

---

183 Van Over, *Eastern Mysticism,* p. 113, n. 3.

184 *Brihadaranyaka Upanishad.*

185 See von Franz, in *Archetypal Dimensions of the Psyche*, pp. 41f.

Buddhism: The Bee and Buddha Nature

Lama Choedak Yuthok, a Tibetan Buddhist Lama living and teaching in Australia, writes that in the Mahayana Buddhist tradition there are several examples given to illustrate the meaning of Buddha nature.[186] First there is analogy of that beautiful, natural mandala, the lotus flower, and then, he says:

> *The second example of Buddha nature is that of bees, which symbolise our anger and agitation. Bees make and accumulate honey for themselves, not for human consumption, so they become agitated and sting us when we take the honey from them. Likewise when others try to take things from us that we value, we become agitated, angry and 'sting' them with our anger. Just as a bee feels it has the right to sting those who steal honey from them, we also feel righteous anger towards those who have taken the things we accumulated through our craving. The more craving we have for the things in our life—wealth, possessions, relationships, ideas and concepts—the more inclined we will be to have anger. Further when we are trying to achieve something, we are agitated and angered by even the slightest mistake we make. This too is like disturbing a hive of bees and we feel the sting of our own anger. What we cannot see is that the thing that stings (i.e. the bee) is the same thing that produces the honey. The honey symbolises our patience: just as the sting of the bee does not affect the honey the bee produces, our anger does not affect the patience that is part of the natural purity of our minds. To endure the impact of our own mistakes or those of others, we must nurture the honey of our patience.*[187]

****

In ancient Tibetan, Indian and also Chinese Buddhism there was the idea that the religious practice of meditation, a form of circumambulation around a sacred centre, serves the goal of producing, within the still living and mortal body, the diamond body into which you move, so to speak. Already in this lifetime you use your diamond body more and more as a dwelling place, so that at the moment of death, like a skin that

---

[186] Lama Choedak Yuthok, *Healing Relationships: Buddhist Insights for Enduring Happiness,* pp. 7f.

[187] Von Franz, *Archetypal Dimensions of the Psyche*, pp. 7f.

falls away from a fruit, this mortal body falls away and the glorified body, a sort of immortal substance as carrier of the individual personality, is already produced by religious practice during a lifetime. This same idea features strongly in alchemical philosophy (see below in chapter five, 'Alchemy and the Individuation Process').

****

It is also true to say that beekeeping is an art that requires a religious attitude. The spirit and mystique of the hive and the bees as well as the miracles of the flowers and fruit, and of the nectar, pollen and honey are ever-present in the task of tending or keeping bees. Sir Edmund Hillary, who with his Sherpa guide Tensing Norgay was the first mountaineer to reach the summit of Mount Everest (Qomolangma), the highest mountain in the world, lived in New Zealand and was a beekeeper. The fact that he was a professional apiarist is often mentioned as a real plus in his embarking on such a journey. This particular summit is regarded not only as the place of the gods but also as a god itself by the local Nepalese and Tibetans—and Hillary and Norgay respected this. On the summit they left a crucifix for Colonel Hunt, the expedition leader, and Norgay, a devout Buddhist, buried some biscuits and chocolate as an offering (of sweetness) to the gods of the mountain. Then they ate a mint cake, strapped on their oxygen tanks and began the climb down.[188]

For humankind, honey as a golden treasure, the sweetness of life, and as healing nourishment has since ancient times been seen as something to quest for, not unlike a Holy Grail. Art from ancient Egypt, for example, bears witness to this, as does the rock art of Australian Aboriginals and the central Indian example shown in Figure 23. One of the earliest images from a rock-art painting at Barranc Fondo, Castellón in eastern Spain, shows honey gatherers (from between 4500 and 4000 B.C.) climbing high on ladders to rob honey from wild hives in rock faces. A similar practice still exists in Nepal.

---

188 Hillary died at the age of 88 on 11 January 2008. See John Elder, 'Death of a Hero: Everest's Great Conqueror', in the Melbourne *Age*, 12 January 2008, 'Insight', p. 3.

Figure 23. Post-Mesolithic rock art, painted in white at Rajat Prapat, central India, showing hanging natural *Apis dorsata* nests and four honey-hunters, some carrying a container for the harvested honey.

4
The Bee, the Honey and the Hive: A Psychological Perspective

Love never settles on that which does not flower or has lost its flower, be it body, soul, or whatever.
—Plato, *Symposium.*

He is not worthy of the honeycomb, that shuns the hives because the bees have stings.
—Shakespeare.

I chose to begin this work or pilgrimage because I was a beekeeper for ten years, had a love for the bees, and also because I was very touched by what Marie-Louise von Franz said in her 1983 lecture, 'The Transformed Berserker: The Union of Psychic Opposites', about honey and its connection with deep love.[189] With this kind of love, a strong fighting spirit and a powerful sting are present.[190] The title suggests the particular opposites evoked. Those of the Berserker, the noble individual who, when it is called for and to help heal a collective situation, can don the pagan bearskin embodying a centred or holy rage—fierce and natural, but at the same time a sacred animal instinct behind the human warrior instinct. This wild but sacred and measured way of approaching the centre or the still point (like the calm 'eye of the storm' or swarm) is a key aspect of what the Berserker symbolises.

A *transformed* Berserker does not express blind, destructive undifferentiated anger and rage but rather a divine fighting spirit or way of

---

189 The lecture was the keynote address at the Eighth Conference of the International Transpersonal Association of Individual Transformation and Universal Responsibility, 27 August to 2 September 1983, in Davos, Switzerland. Another version of this talk appeared in *ReVision* vol. 8, no. 1 (Summer/Fall 1985), and key excerpts are included in her book, *Archetypal Dimensions of the Psyche.*

190 Ibid., pp. 36f.

protecting and bringing balance to what is warring or in opposition in the collective. The bee, although a tiny insect, has unique qualities that help us amplify and thus understand the nature of both the Berserker and the transformed Berserker, which I will discuss.

We already have some insight from Persia in the story told by the poet Nizami of the bee that carried crumbs from the Sultan's table to feed a blind sparrow:

> *A hornet seized a bee, intending to devour it. The bee begged for its life and said. "My hive contains so much honey, I myself am of so little worth, why do you not go to the hive and leave me in peace?" The hornet replied, "The hive is sweet because of the honey in it, but you must be sweeter as you are the source and spring of the honey."*[191]

The Opposites of Love and War

In mythology and religion of many ancient cultures, several significant Aphrodite figures like Astarte and Inanna are goddesses of both love and war. They are not only *love* goddesses. The warring chthonic part of the feminine nature and embodied sexuality needs to be recognised and have a place in love. Jung, in writing of sexuality, says:

> *Sexuality is of the greatest importance in the expression of the chthonic spirit. That spirit is 'the other face of god', the dark side of the God-image. The question of the chthonic spirit has occupied me ever since I began to delve into the world of alchemy.*[192]

The sexuality of bees is in itself a mystery as the asexual worker female can perform miracles if there is a need to save the hive. But, in general, the sexual element in worker bees could be explained as being more about that of the flowers, as Rudolf Steiner said in a lecture on bees in 1923. He referred to an old wisdom:

> *[that the] wonderful activity of the beehive pointed back to the condition of the bees' love life that these people [living in antiquity] associated in their mind with the plant and with the goddess Venus-Aphrodite. The whole beehive is permeated with life based on love. In many ways the bees*

---

191 Ransome, *The Sacred Bee,* p. 74.

192 *Memories, Dreams, Reflections,* p. 192.

renounce love. And thereby this love develops within the entire beehive. You'll begin to understand the life of bees once you are clear that the bee lives as if it were in an atmosphere pervaded thoroughly by love.

But the thing a bee profits from the most is that it derives its substance from the very parts of a plant that are pervaded by a plant's love life. The bee sucks its nourishment, which it makes into honey, from the parts of a plant that are steeped in love life. And the bee, if one may express it this way, brings love from the flowers into the beehive. So it becomes clear that one needs to study the life of bees from the standpoint of the soul.

The bee, with the exception of the queen bee, might say, if I may put it this way: "As individuals we want to renounce all sexual life, so that we make each one of us into a supporter of the hive's love life." They have indeed carried into the hive that which lives in the flowers. When you begin to think through all of this properly, you will have unlocked the whole secret of the beehive. The living element of this thriving, germinating love that is spread out over the flowers is also contained in the honey that the bees make.[193]

Bee symbolism also combines the opposites, particularly of the honey and the sting. This is a key part of my topic. Where there is love or passion there is the possibility of experiencing its opposite—hate or a wild, fighting spirit, or even death, as is indicated by the religious use of the word passion. Sylvia Plath's bee poem 'Stings',[194] for example, is both a love poem and a war poem about her struggle with her husband Ted Hughes, as well as part of an attempt to find a new sense of centring and self.[195] Ted Hughes's poem 'The Bee God', written many years later, powerfully describes this 'same' event where he was badly stung by many bees. Their relationship was a fierce love story and their poetry was very much part of their individuation processes.[196]

Bees are strongly associated with the archetype of order and with virginity. This links them with the qualities of the astrological sign of Virgo and encompasses a sacred sexuality, a *hierosgamos*—as Jung said, a

---

193 Steiner, *Bees*, p. 3.

194 Plath, in *Ariel,* p. 60.

195 Sylvia Plath had read Jung, as had her estranged husband Ted Hughes.

196 Hughes, in *Birthday Letters,* p.150.

union 'more primitive, more primeval and more spiritual, than anything we can describe'.[197]

Jungian analyst/astrologer Liz Greene in her description of Virgo mentions the meting out of justice.[198] The goddess Astraea (or Dike) represents the principle of justice and is usually depicted carrying a sheaf of barley. According to Hesiod she was the daughter of Zeus. Jane Harrison, in her study of the social origins of Greek religions, writes:

> *Dike [Astraea] is the way of life of each natural thing, each plant, each animal, each man. It is also the way, the usage, the regular course of that great animal the Universe, the way that is made manifest in the Seasons, in the life and death of vegetation; and when it comes to be seen that these depend on the heavenly bodies, Dike is manifest in the changes of the rising and setting of constellations, in the waxing and waning of the Moon and in the daily and yearly course of the Sun.*[199]

In the natural zodiac Virgo is the sixth house, the one that is just below the horizon. It represents the daily rituals and also the sacred healing potential of just being (or 'be[e]-ing?' as a colleague suggested when she learnt I was writing about bees—that is, *being* as opposed to doing). This is the zen-like daily routine, like that of the bees and their methodical dedicated work amongst the flowers and in the hive. The bees, like those related to something infinite in their daily lives (for example, to the inner god-image, the psychological Self), have the sting and passion to protect what protects them.

> *Understand ye Sons of wisdom, the Stone declares: protect me, and I will protect thee: give me my own, that I may help thee.*[200]

Thus, the bees look after the hive with its focus on the queen and preserving life. They protect, and in return are protected.

A hive of bees does not need to be watched and directed by humans. The individual bees and the hive have their own inherent momentum,

---

197 Letter to Mary Mellon, 18 April 1941, in *C. G. Jung Letters*, vol. 1, p. 298.

198 *The Astrology of Fate,* p. 212.

199 Harrison, *Themis*, p. 517.

200 From 'The Golden Treatise of Hermes', in M. A. Atwood, *Hermetic Philosophy and Alchemy,* p. 128.

their own daily and seasonal rhythms. In the *Aeneid,* Virgil describes how the bees work and build as a metaphor for the great feat of building Carthage:

> *As when in the early summer o'er the flowery mead beneath the sun the honey bees are at work, when they lead forth the full-grown young of their race, or when they pack the liquid honey and fill their combs full of sweet nectar, or receive the burdens of those that come in, or form a martial line and drive forth from the hives the drones, a lazy tribe; hotly glows the work, and the sweet-scented honey is fragrant with thyme.*[201]

The alchemical work and the psychological individuation process involve hard work and daily routine as in the stage in alchemy called 'women washing,' with its images of women doing the laundry. This is the humble repetition of the necessary daily tasks of life—the down-to-earth stage.[202] Marie-Louise von Franz in her lectures on alchemy comments on this stage:

> *In alchemical literature it is generally said the great effort and trouble continues from the nigredo to the albedo; that is said to be the hard part, and afterwards everything becomes easier. The nigredo—the blackness, the terrible depression and state of dissolution—has to be compensated by the hard work of the alchemist and that hard work consists, among other things, in constant washing; therefore even the work of washerwomen is often mentioned in the texts.*[203]

If bees were not industrious, we would not have honey. Von Franz notes that when we are in the grip of an élan vital, doing creative work or getting ready to do it, this can be a difficult time of impasse. But she adds:

> *One needs to be carried by the enthusiasm of the big goal: otherwise one would give up and prefer to live a comfortable life. One must never forget that man's greatest passion, greater than any other, is probably laziness,*

---

201 Virgil, *Aeneid,* I, 430f.

202 As in the sixteenth-century alchemical text by S. Trismosin, *The Splendor Solis;* see J. Henderson and D. Sherwood, *Transformation of the Psyche: The Symbolic Alchemy of the* Splendor Solis, p. 167.

203 Von Franz, *Alchemy*, p. 220.

and to overcome the passion of laziness one needs a lot of energy, characterised by the great and Titanic formations.[204]

She speaks of the galvanising energy of 'holy rage', which I shall later discuss further in relation to the Transformed Berserker vision, and she gives a strong example of how Jung struggled to write *Psychological Types* until helped by the forces of the following dream:

He [Jung] saw in a port an enormous ship laden with goods, which should now be pulled into port. Attached to the ship on a cord was a very elegant white horse, absolutely incapable of pulling in the ship, which was much too heavy! At this moment there came suddenly through the crowd an enormous red-headed giant who pushed everybody aside, took an axe and killed the white horse, took the rope and with one pull, pulled the whole ship into port.[205]

It is the living of everyday life that allows us to keep our feet on the ground and feel stable and centred. Jung, when he was going through his crisis in 1913 and was courageously and actively engaging with the images from the unconscious, emphasised in his autobiography that it was important to be able to orient or ground himself like this. He wrote:

Particularly at this time when I was working on the fantasies, I needed a point of support in 'this world', and I may say my family and my professional work were that to me. It was essential for me to have a normal life in the real world as a counterpoise to that strange inner world. My family and my profession remained the base to which I could always return, assuring me that I was an actual existing, ordinary person. The unconscious contents could have driven me out of my wits. But my family, and the knowledge [that] I have a medical diploma from a Swiss university, I have a wife and five children, I live at 228 Seestrasse in Küsnacht—these were the actualities which made demands upon me and proved to me again and again that I really existed.[206]

---

204 Von Franz, *Creation Myths,* p. 179. She also speaks of laziness in *The Feminine in Fairy Tales*: 'Most men with a positive mother complex are lazy, for the mother is the symbol of matter and matter is inertia. The positive mother is like a big feather coverlet which always blocks the man who will naturally tend to be lazy.' (pp. 62f.)

205 *Creation Myths*, p. 173.

206 *Memories, Dreams, Reflections,* p. 214.

In other words, when we are feeling the most psychologically disoriented, the need to attend to or be cognisant of everyday practicalities and responsibilities (like housework, eating, washing, paying bills) provides a support structure. A fundamental question that we can continue to ask ourselves each day is: What has to be welcomed and accommodated in ordinary life so that order, rather than destructive or overwhelming chaos or illness manifests? How can I be centred?

The ritual of doing housework is only one of those things. Busyness and house or 'temple' cleaning tasks are important to the worker bees, who are particular about maintaining hygiene and order in the hive. Much work goes into producing the honey—so much life force comes from the plant and animal realms, from the sun and the earth.

Alchemy, like the work of the bees, finds itself involved in the humble, everyday tasks of life, and in hard work and repetitive tasks. As epitomised in the famous Zen Buddhist saying: 'Chop wood, carry water', such everyday essential work is a form of participation in the miraculous transformations of nature and of being in healing contact with the earth. Thus it is a positive way[207] of being in contact with the Great Mother and so with her fertility and a sense of lunar waxing and waning, and the cyclical flow of the indestructible life force—the round of birth, death and regeneration. Eros in itself contains its opposites. As Jung said,

> *Eros is a questionable fellow and will always remain so... He belongs on one side to man's primordial animal nature which will endure as long as man has an animal body. On the other side he is related to the highest forms of the spirit. But he thrives only when spirit and instinct are in right harmony'.*[208]

For example, the honey-making and the stinging aspects of the bee, belong together—as alchemist Michael Maier wrote: 'Where honey is, there gall [bitterness]'.[209] Honey is linked with the sweetness of life (elixir) and with love. It is an elixir, and an ancient healing and pure food. It is made from what the worker bees gather from the flowers and

---

207 Or the way of Adonis or Attis.

208 'The Eros Theory', *Two Essays on Analytical Psychology*, CW 7, par. 32.

209 Michael Maier, in *Symbola Aurea Mensae Duodecim Nationum*, p. 515.

bring back to the hive to deposit in the hexagonal honey cells that make up the honeycomb: the basic but miraculous structure of the hive.

Wounding and Healing: Cupid's Arrows, the Bee's Sting and Honey

The honey, the quintessential sweet substance, and the sting are extreme opposites. The sting kills the stinger (that is, the bee) and can cause great discomfort and, in some cases, death to the one who is stung. The sting is arrow shaped, as noted earlier.

What does an arrow do? It pierces the skin or flesh in such a way that it can do much damage if not carefully removed. The arrowhead works like a hook that can tear the flesh apart and inflict a terrible wound. Removing a bee sting is a sensitive process too. An arrow tears the flesh apart and creates an open wound. It is not just a surface wound; we are pierced: things go deeper. A bee's sting is shaped like an arrow and works on the same principle (see the fine anatomical drawings by Jan Swammerdam in Figure 24, next page).

Consider for example the various art works that depict Cupid or Eros and his mother Aphrodite. Cupid is sometimes shown with a bow and arrows that wound by piercing the body, specifically the heart, opening it up to the possibility of a new love, but at the same time inflicting a dreadful, tearing wound. For good measure, arrows in mythology were mostly dipped in a strong substance (like the blood of the Hydra) that was usually deadly poisonous, if it were not instead taken in a small homeopathic dose as a healing tincture. A homeopathic measure respects the essence of the substance and administers it in a minute but effective dose.

Several poets of ancient Greece have written of Eros, the god of love (Cupid to the Romans), complaining bitterly to his mother–lover Aphrodite (Venus) when stung by a bee. The following poem is by an unknown imitator of Theocritus, and is called 'The Robber of Honeycombs'.

> *The thievish Love—a cruel bee once stung him as he was rifling honey from the hives, and pricked his fingertips all: then he was in pain, and blew upon his hand and leaped and stamped upon the ground, and he showed his hurt to Aphrodite and made much complaint, how the bee is a little creature and yet what wounds it deals! And his mother laughed and*

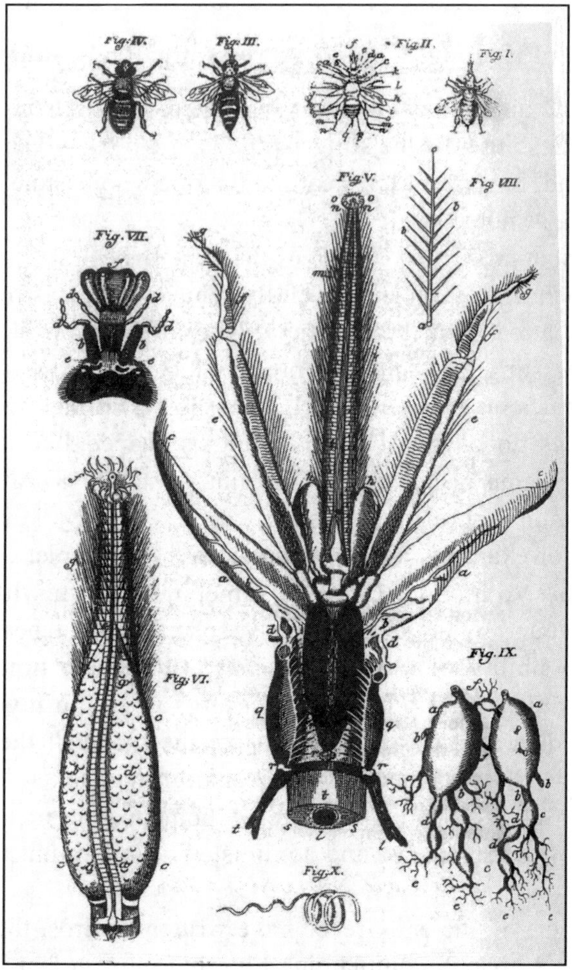

Figure 24. Drawings of bee anatomy by Jan Swammerdam (b. 1637), featuring the sting; from *The Book of Nature* (1737).

said, 'Art thou not even such a creature as the bees, for tiny thou art, but what wounds thou dealest!' [210]

Jung talks about the piercing arrow in several deeply transformative archetypal contexts. Adam (the transformative substance, the old Adam who is to renew himself) is addressed by an unknown being: 'Thus I step up to you with my sharpened arrows, aiming them at your heart'.[211] Jung says these arrows recall the passion of Mercurius (*telum passion*—not unlike Cupid's arrows) and the shafts of Luna, which the alchemists link to the Song of Songs—'Thou has wounded my heart'.[212] I shall return to this theme in chapter five, 'Alchemy and the Individuation Process'.

Questions of acceptance of the opposites and the transforming of the sting and considerations of healing lead us now to further discussion of the Transformed Berserker vision of Niklaus von Flüe.

The Transformed Berserker

As I have mentioned, I chose to begin this work or pilgrimage (a *perigrinatio animae*) because I was very touched by what Marie-Louise von Franz said about honey and its connection with deep love in her lecture on 'The Transformed Berserker'.[213]

Jung is often associated with the idea that where there is love, power issues and struggles are not present; and that where power dominates, there is no love; for instance:

> Where love reigns there is no will to power; and where the will to power is paramount, love is lacking. The one is but the shadow of the other: the man who adopts the standpoint of Eros finds his compensatory opposite in the will to power, and that of the man who puts the accent on power is Eros.[214]

In the fighting spirit of the bear-like Berserker, or of the bees or of Cupid/Eros, the god of love, there is a force or stinging that is present but

---

210 Theocritus, XIX ('Love Stealing Honey').

211 *Mysterium Coniunctionis,* CW 14, par. 549.

212 Ibid., par. 24; see also par. 25.

213 See above, p. 112.

214 'The Problem of the Attitude-Type', *Two Essays*, CW 7, par. 78.

that can be transformed symbolically by deep relatedness or eros. Love and war, war and peace, eros and thanatos exist together. Central to this theme, in von Franz's lecture on 'The Transformed Berserker', is a powerful and inherently healing vision of the fifteenth-century Swiss hermit and saint Niklaus von Flüe, who after midlife became profoundly introverted.

Like the rainmaker in the story of the Rainmaker of Kiaochau,[215] Niklaus was able, through spiritual devotion to his perfection and process of transformation, to exercise political influence on the society and environment around him, in the manner of an eastern sage, helping heaven and earth come together again after a separation, by bringing his own inner heaven and earth into harmony.[216] This healing harmony came about through long periods of prayer or a circumambulation of the centre or Self—an example of the alchemical process of *circulatio* as well as *multiplicatio*, which I will discuss here later.

Marie-Louise von Franz explores the question of what underlay the extraordinary effect that Niklaus von Flüe had on those around him. Her starting point was Brother Klaus's life-changing vision:

> *It seemed to Brother Klaus that a man who looked like a pilgrim was coming towards him. This man held a staff in his hands, had on a hat with the brim turned back in the manner of a wayfarer, and wore a cloak, Klaus knew within himself that this man came from the east or from far away. Although the pilgrim did not say so, Klaus knew he came from "where the sun rises in summertime". He stood in front of Klaus singing the word "Hallelujah!" When he began to sing, his voice echoed, and everything between heaven and earth seemed to support his voice. And Klaus heard "three perfect words that stood out from the rest, coming out of one origin", which then closed up again like something on a spring. When Klaus heard these three perfect words, none of which touched either of the others, they nonetheless struck him as being a single word. When the pilgrim had finished his song, he asked Klaus for alms. Brother Klaus suddenly*

---

215 It is interesting to note that bees are attributed with the ability of foresight and of making rain in the folk tales of various cultures. For the Rainmaker of Kiaochau story, see *Mysterium Coniunctionis,* CW 14, par. 604, n. 211.

216 Ibid., par. 604.

had a penny in his hand and dropped it into the pilgrim's hat. And so the man [Brother Klaus?] had never realised that it was a thing so worthy of veneration to receive a gift in one's hat.

Klaus asked where the wayfarer came from and who he was, and the traveller said only, "I come from there", and was unwilling to say anything further. Klaus stood in front of him and looked at him. Then the pilgrim transformed. He now no longer wore a hat and cloak, but rather a blue-grey vest. He was a fine, handsome-looking man, and Klaus looked at him, with joy and longing. The brownish colour of [the wayfarer's] face gave him a noble look; his eyes were black like a magnet, and his limbs of extraordinary beauty. Although he was clothed, Klaus could see his limbs. As Klaus was looking at him so raptly, the wayfarer also looked back at him. In this moment, great miracles occurred: Mount Pilatus collapsed to the ground and was completely flat; the earth opened up; Klaus thought he could see the sins of the whole world. A huge throng of people appeared to him, and behind them appeared the truth, but all the people had turned their backs on it. In their hearts, Klaus saw a great sickness, a tumour as big as two fists. This sickness was egotism, by which people were so seduced that they were unable to bear the sight of the man (of truth), "no more than people can stand fire". In great confusion, fear, and shame, they ran hither and thither, and finally fled; "but the truth remained there".

Then the countenance of the wayfarer transformed "like a veronica", and Klaus had a great longing to see more of him. He [Klaus] saw him again as before, but his clothing had changed, and he stood before him in a bearskin with coat and pants. The fur was spangled with golden colour, but Klaus saw clearly that it was a bearskin. The bearskin was very becoming to the pilgrim, and Klaus recognised his extraordinary beauty. As he stood before the wayfarer, so noble in the bearskin, Klaus saw that the figure wished to bid him farewell. Klaus asked him, "Where do you want to go?" and he replied, "I want to go up country" and would say no more. As the pilgrim departed, Klaus stared after him and saw that the bearskin shone on him as when someone moves a brightly polished sword back and forth and its reflection is seen on a wall. And Klaus thought that this was something whose meaning would remain hidden from him. When the wayfarer had gone maybe four steps, he turned around, took off his hat, and bowed to Klaus. Then Klaus realised that the wayfarer bore him such love that he was quite stricken and had to confess that he was not deserving of this great love. Then he saw that this love was in the wayfarer. And he saw that his spirit, his face, his eyes, his whole body was full of this elevated love (Minne), *like a vessel that is filled to the brim with honey. Then he could no longer see the wayfarer, but he was so fulfilled that he no longer desired anything from him. It seemed to Klaus that the wayfarer had revealed to him "everything that was in heaven and on the earth".*[217]

There are various transformations in this vision. The first to appear is the unknown wayfarer or god figure, wearing blue–grey and the cloak and hat of Wotan, the shamanistic Northern Germanic god of war, the storm and the 'wild hunt'—a god of revelation who moves tirelessly around the countryside like Hermes/Mercury. When the pilgrim sings the sacred word of praise, there is a response from the whole cosmos. It is as if the fruit of a fertilising unity of consciousness in relationship with the unconscious is witness to the upwelling of an openhearted receptiveness to the flow of eros. The darker, more occluded, god figure then wears the bearskin, transforms and becomes golden and radiates light (consciousness) and drips honey (eros or love). Spirit and matter, the above and below, are united, as in alchemy, in a middle place. And as we have seen, the middle place is also the place of the bee, the honey-maker.

The Transformed Berserker is a numinous symbol of differentiated eros, the kind that emanates a fructifying, healing influence on warring elements in the collective. An outer crisis and collapse is mediated in the vision and this has a wider effect on the objective world, while at the same time revealing, as von Franz says, the spiritual inner world of man. The pilgrim in Klaus's vision, she says, could be interpreted as a 'spirit of truth', one whose effect is more connected with the embodied passion of being stirred deeply and emotionally—more so than is usually associated with the descent of the Holy Spirit inspiring humankind 'from above'. The deeply felt and transformative eros in the vision, instead arises viscerally from the depths of the instincts. It is an embodied and magnetic attraction—a passionate love. And, as von Franz says, this 'is, indeed, the secret of love itself'.[218]

---

217 Von Franz, *Archetypal Dimensions of the Psyche,* pp. 39f (modified in translation).

218 See von Franz's work on the visions of Niklaus von Flüe (currently only available in German in book form and on DVD with an English language option).

Brother Klaus, at the time of this vision, was extremely troubled over the social and political injustice and cruelty taking place around him. Von Franz says it seems that this Berserker (this Wotanic figure wearing a bearskin)[219] is overflowing with honey, that is with love as a compensating force. In psychological terms, this suggests that brutality, if integrated (illuminating the inner darkness with consciousness), would not then operate autonomously in a way that manifests as violence in the outer world.

****

Among the followers of Wotan, there were the 'raging ones' (the Berserkers; *Ber* = bear; *serk* = skin). Von Franz describes how this capacity to 'go berserk' (in Old Norse meaning literally to 'change one's coverings', as we see in the vision) was thought to be a gift in certain noble families. 'If a battle was taking place and the Duke or Count sat at home', she said, 'suddenly he would give a terrific yawn and then fall into a deathlike sleep, while on the battlefield there appeared a bear who killed everybody.' After a while the bear disappeared and the Duke woke up and was very tired. He had 'gone berserk' and in the form of an exteriorised soul, a bear's soul, had fought the battle. These bears were looked on as ghost-bears who did great deeds, and the proof that it really was the Duke himself was that the bear was wounded in the right paw, and that the man when he woke up at home was wounded in his right hand, for example.[220]

The Berserker in the vision is a numinous man with theriomorphic attributes since, as Jung commented, 'he surpasses the ordinary man not only upwards, but downwards'.[221] Jung, in his letter to Professor Blanke in 1945, says that the Wotan image in the vision indicates Brother

---

219 Other details regarding the sacred significance of the bear include the following. In warmer parts of southern Europe, the Great Mother Artemis was known as the Great She-Bear, Ursa Major, seeker of the honey of the goddess, and of introspection and feminine receptive energy. In the colder north, for example in Lapland, the bear is called *siavo;* 'holy', or 'the holy beast'. And as 'Corn Bear', the bear was a symbol for vegetation and fertility and, as such, played a role in Germanic marriage rites.

220 Von Franz, *Feminine in Fairy Tales*, pp. 54f.

221 *Letters*, vol. 1: letter to Fritz Blanke of 2 May 1945, pp. 363f.

Klaus's Germanic 'mentality' and that the Christ or Self figures that the wayfarer transforms into are, firstly, a pilgrim who like the mystic (and alchemist searching the world for the phoenix feather in Michael Maier's *peregrinatio* text) has gone on the *peregrinatio animae;* and, secondly, a bear (an animal that in alchemy is a symbol of transformation) whose pelt has a golden lustre indicating the *novus sol* (new sun) or living gold. It is interesting to also note that the bear is at the top of the list of the bee's natural enemies. Figure 25 gives a hint about why this is so. Bears love honey and to reach it they need to overcome the stinging aspect of the bees who fiercely guard their treasure.

Figure 25. A European brown bear drowning bees before taking their honey-combs. Engraving from a Spanish book of emblems *(Saavedra Fajardo,* 1640).

Early in the vision, Klaus looked for a moment into hell and saw the ills of the world like a dark tumour. And truth was relegated to the background. If we can really confront the unconscious by taking inner figures and everything that they have to communicate to us seriously (including the brutal shadow side), then we are open to a more related connection (a spiritual eros) with the eternal inner god-image. After Klaus was able to look at its opposite, deep love was constellated and became visible as a golden, flowing radiance of light and honey. This von Franz and Jung describe as the 'unalloyed man'. The transformed Berserker sparkles with golden light as if overflowing with honey. A multiplying and intensifying healing effect emanates from this source or centre.

Often something of the essence of the Self is seen or revealed in the way a person approaches a deep crisis, by the quality of his or her response to a sense of some lack or defeat. A fighting spirit can come from the centre of our personality and be sagacious rather than bellicose or destructive. The mystery of individual transformation comes through just being, through living our lives authentically and remaining openhearted to the experience of meeting and responding with eros to an unknown visitor or wayfarer. It is possible that this will be a god walking on earth. However, with every transformation or change of condition there is a time (an interim or interregnum period) of disorientation—of swarming—that, as we have seen in the life of the bees, is the ground for the new life or renewal of the mystery of the queen. We take a new step and the former equilibrium is now gone and is no longer with us. This is a kind of crisis, often accompanied by a sense of partial failure or defeat.

Work, employment or service, and real lived-experience are usually what are needed for a disciple, apprentice or initiate to attain to mastery (as this is something which has to be earned). There is a daily round of ritual or everyday tasks to attend to, often for maintaining health and order in the collective as well as the individual. The inner self has somehow been aroused by the pending change; instinctual urges and desires have to be transformed if one wishes to contribute to the productivity and growth of the community (in the case of the bees, this is to the hive), a form of *ecclesia spiritualis*.

The life of the bee as a symbol for the individuation process involves, as we have seen, a consideration of individual transformation and of social or collective responsibility, bridging spirit and matter. The alchemical transformation process or personal individuation involves the following development, as Theodor Abt describes it:

> *The aim of alchemy is that the adept learns during his or her lifetime, a way to relate to an inner centre or self within his or her own psyche. This inner centre corresponds to the God-image of all religions. It is symbolised by the stone of the sages or the anthropos that is first buried in the unconscious, and only when the individual continuously relates to it and works on its images does it turn into a reliable psychopompos (inner psychic guide). This is what we would call today with C. G. Jung the realisation of the self.*[222]

The pilgrim in the vision of Brother Klaus clearly can be seen as an image of the Self. Von Franz describes this wonderfully as Klaus's eternal, inner spiritual core, something like the inner Christ as described by the mystics. This unknown, mysterious man who walks the earth and meets and then greets Niklaus is dressed in the blue–grey cape and hat of Wotan or Odin as wayfarer, and has the stature and presence of this pagan god. Wotan is not only a passionate god of the storm but also of deep eros—as a god-image he combines and holds the tension of the opposites of love and war. He is also a figure who bridges the natural and divine forces of sky and earth as a seer and shaman, linking the here and now with the beyond. Brother Klaus's unconscious produced a Christ figure, which like the philosopher's *lapis* unified the opposites. The dark side of this figure, which looks back to the pagan Germanic tradition, is a Berserker. The aggression of the Berserker is not autonomous, however, but is integrated into the inner wholeness of man's Berserker shadow.[223]

An important point to conclude with is that the alchemists' description

---

222 T. Abt, *The Book of Pictures:* Mushaf as-suwar *by Zosimos of Panopolis,* p. 39.

223 Von Franz, *Archetypal Dimensions of the Psyche,* p. 48. The atmosphere that Wotan creates unlocks feelings, poetic ecstasy and clairvoyant trances. Jung says of him: 'He is the god of the storm and frenzy, the unleasher of passions and the lust of battle; moreover he is a superlative magician and artist in illusion who is versed in all secrets of an occult nature.' ('Wotan', *Civilisation in Transition*, CW 10, par. 375)

of the *lapis* as a living flowing stone is re-evoked here by the archetypal image of the 'cosmic Berserker overflowing with honey'.[224] This is the living stone that streams forth (like the light and honey from Zeus's birth cave) 'rose-coloured blood' (which is described as being more like nectar than blood) or a rose colour, and has a healing effect on its environment.

Both Jung and von Franz remind us of the redemption and healing in the alchemist Gerhard Dorn's description of the stone of the sages. Dorn says:

> *They* [the philosophers or sages] *call their stone animate because at the final operations,* [it] *sweats out drop by drop . . . a dark red liquid* [elsewhere described as pink and like nectar] *like blood. . . . And for this reason they have prophesied that in the last days a most pure man* [putus = genuine, not falsified, i.e., unalloyed], *through whom the world will be freed, will come to earth and will sweat bloody drops of a rosy or red hue, whereby the world will be redeemed from its Fall. In like manner, too, the blood of their stone will free the leprous metals and also men from their diseases . . . For in the blood of this stone is hidden its soul.*[225]

Like with the beehive, as a living mysterious centre from which the life of the bees radiates and flows, producing a golden, dripping healing elixir, the stone of the sages lives and flows and radiates its unalloyed healing influence on both microcosm and macrocosm.

---

224 Von Franz, ibid., p. 49.

225 *Alchemical Studies*, CW 13, par. 381.

5
Alchemy and the Individuation Process

The psychological individuation process, like the bees' honey making, is an *opus magnum;* and the hive of bees in some ways symbolises a community of the Self (the *ecclesia spiritualis* understood alchemically as those who take part in the transformation and union of opposites).[226] The vitality of the whole hive or community, and thus the fertility of the queen, is something that all the bees are instinctively motivated by. This is like a guiding principle of *telos* or *zoë*—of life going on and renewing itself, like water gushing up from a deep underground spring or an eternal source.

We have also seen that, in the symbolism of the bee, there is a strong theme of the meeting of the above and the below, and an alchemical style of transformation of matter into an eternal substance or healing elixir—the honey. It is helpful to consider the psychological *opus magnum* (the individuation process) and the alchemical *lapis*-making or *chrysopoeia* (the gold-making), in the light of the bees' honey making and their circumambulation of the centres of the flowers, the hive and the queen. And in doing this, one needs also to consider the symbolism of pollination and cross-pollination—of the blossoming followed by the *coagulatio* of the fruit that often symbolises the goal of the alchemical *opus* and psychological individuation process.

The goal of the individuation process is to achieve a conscious relationship with the Self. The goal of the alchemical procedure was most frequently represented by the philosopher's stone, the *lapis*.

Psychologically, developing an enduring inner centre is the goal and everything is directed towards and circumambulates that centre, the Self.[227] The life of the bees revolves around those centres of the flowers outside the hive and the queen inside. The process of circumambulation

---

226 Von Franz, 'Commentary', in *Aurora Consurgens*, p. 403.

227 Jung, *Memories, Dreams, Reflections,* p. 224.

is the quintessence of the life of the bees and the hive, and the famous dance language is an important variation on this theme.

As we have seen, bees are deeply entrenched in the mythos of many cultures worldwide. As nature's pollinators they are essential to all life. On the macrocosmic level they are faithful servants of the Great Mother and Mother Nature, and this is exemplified on a microcosmic level in individual colonies through dedication to their queen. Bees are icons of androgyny and, along with their delicious healing and preserving honey, feature strongly in the symbolism of the erotic and sensual; and of love, death and immortality. They are also depicted as journeyers or wayfarers and divine emissaries and transformers like Mercurius and Wotan, the Delphic Bee and Proteus. However, we need always to remember that with bees there is a stinging and a swarming side—a wild Berserker shadow and even a slothful, vegetative drone aspect that needs to be recognised and respected.

Jung, in noting the importance of the psychological opposites, writes at the beginning of *Mysterium Coniunctionis: An Inquiry into the Separation and Synthesis of Psychic Opposites in Alchemy:*

> *The pairs of opposites constitute the phenomenology of the paradoxical self, man's totality. That is why their symbolism makes use of cosmic expressions like coelum/ terra. The intensity of the conflict is expressed in symbols like fire and water, height and depth, life and death.*[228]

The Alchemical Stages of Making the *Lapis*

Marie-Louise von Franz describes the process, and work of preparation, that is undertaken or lived through before the final red–gold stage (the *rubedo)* of *lapis*-making is reached. The black stage (the *nigredo)* is one of descent into the unconscious. One's old conscious attitude is no longer lighting the way. This, as shown above in the 'Wintering' section (chapter one), can be a time of darkness and depression. Then a 'greenness' of flowers and animals might be described, and vitality and light returns in the white stage (the *albedo).* Next comes the difficult, but important, stage of engaging with the anima and animus 'problem'. Introversion,

---

228 *Mysterium Coniunctionis,* CW 14, par. 4.

much reflection and a well-sealed vessel are needed to contain this potent stage of the alchemical and psychological transformation.

Von Franz emphasises working on the anima and animus in a closed vessel—keeping the retort of one's own inner work closed completely to reflect on oneself in order to become conscious of the difficult anima or animus aspect of the unconscious. She says:

> *In the black and the white stages in alchemy, the alchemist has to work hard, and in analysis the analysand has to work hard. It is an opus to try to become conscious of these powers [the anima and animus]. However, after this part is done the alchemists say the hard work is over. Now one has only to go on and to warm one's substance with a lenient, mild fire, and not to make any more effort, then the rubedo appears. With that one can open the retort and, in alchemical language, let the philosopher's sun or stone come out and take over, as they say, the ruling of the world.*[229]

This, in depth psychological terms, means the beginning of the realisation of the archetype of the Self. The solidity of the vessel has now become the solidity of the philosopher's stone: that is, a permanent inner experience of the Self which now gives the personality substance, 'making an outer artificial solidity or solidification superfluous', as von Franz says.[230]

Jung used the term 'Self' to express the idea of the centre that is also the circumference—the totality of the personality, embracing all (both consciousness and the unconscious). The centre of being has a thousand names, he says: the Atman in India, Christ in Christianity, the stone in alchemy, the diamond, the Divine child, the flower, the circle, the square, the Tao in China. All these are but a few of the symbols through which humans have experienced this central mystery of life.[231]

In alchemy the making of the *lapis* or stone is the same as building up the 'glorified body'[232]—the diamond or resurrection body—that is, 'an immortal nucleus of the personality'. To do this, the alchemists say, you

---

229 Von Franz, *Apuleius' Golden Ass,* X, pp. 8f.

230 Ibid.

231 Helen Luke, *Dark Wood to White Rose*, p. 162.

232 Von Franz, *Creation Myths*, p. 222.

must repeat the whole process of creation, as the work of creating a unique, conscious, whole individual (microcosm) psychologically is symbolically equivalent on a macrocosmic level to the creation of the world. This is a central theme in alchemy, found in many texts including the following example from the 'Emerald Tablet of Hermes':

> *10 Thus the world has been created.*
>
> *11 From here will come marvellous adaptations [miracles], whose manner this is.*[233]

In the Far East, the Golden Flower is a symbol of the Self, a familiar image for the eternal in humans. Buddha is supposed to have once given a wordless sermon during which he presented a yellow flower (or perhaps a white lotus) to his disciples. In the Chinese Taoist text *The Secret of the Golden Flower,* the Golden Flower is described as a 'new being', which, through meditation, unfolds from the dark inner depths into silence. This is what Jung calls the experience of the Self.[234]

The *Lapis* and *Multiplicatio*

Miracles in terms of the alchemical *opus* are often linked with synchronicity and *multiplicatio* at the end of the process. The personal dimension takes on a miraculous quality because it is penetrated by the transcendent dimension. And when one is open to the unconscious, synchronicities happen that correspond to these miracles. (For example, in the last of the famous series of Zen ox-herding pictures: the unassuming sage walks by and the trees blossom; or the rain falls after the rainmaker sits meditating or reflecting in his hut for three days and gets himself 'into tao', in balance with his surroundings.) Thus the idea of the Self, like that of the oriental tao and atman, is at least in part a product of cognition, grounded neither in faith nor in metaphysical speculation but in the experience that under certain circumstances the unconscious spontaneously brings forth an archetypal symbol of wholeness, as we have seen in the Transformed Berserker vision of Niklaus von Flüe.

---

233 Edinger, *The Mysterium Lectures,* p. 160.

234 Von Franz, *On Dreams and Death*, p. 33.

Individuation: The *Opus*

At the beginning of the alchemical process there is the *massa confusa,* not a union of, but a conflict of, the opposites. Jung described the notion of warring opposites and the possibility of union like this:

> *Alchemy, with its wealth of symbols, gives us an insight into an endeavour of the human mind, one which could be compared with a religious rite, an* opus divinum. *The difference between them is that the alchemical opus was not a collective activity rigorously defined as to its form and content, but rather, despite the similarity of their fundamental principles, an individual undertaking, quest or process in which the adept staked his whole soul* [235] *for the transcendental purpose of producing a unity.*[236] *It was a work of reconciliation between apparently incompatible opposites, which, characteristically, were understood not merely as the natural hostility of the physical elements but at the same time as a moral conflict. Since the object of this endeavour was seen outside as well as inside, as both physical and psychic, the work extended as it were through the whole of nature, and its goal consisted in a symbol that had at once an empirical and a transcendental aspect.*[237]

Union: Conjunction

As in the individuation process, the hive, the living community of bees, is a strong example of natural vitality and self-regulation in a 'community of the spirit'. Marie-Louise von Franz, in an essay on the individuation process, wrote the following about the community of the Self:

> *But, as is the case with all inner processes, it is ultimately the Self that orders and regulates one's human relationships, so long as the conscious ego takes the trouble to detect the delusive projections and deals with these inside himself instead of outside. It is in this way that spiritually attuned and similarly oriented people find their way to one another, to create a group that cuts across all the usual social and organisational affili-*

---

[235] Bees sting to protect their queen, honey and hive. When they sting, they die. So for them to sting is an act of self-sacrifice in order to save the whole. Scientists have sought to investigate human altruism through researching bees.

[236] Like the hive of bees protecting their queen.

[237] 'Epilogue', *Mysterium Coniunctionis*, CW 14, par. 790.

ations of people. Such a group is not in conflict with others: it is merely different and independent. The consciously realised process of individuation thus changes a person's relationships. The familiar bonds such as kinship or common interest are replaced by a different type of unity—a bond through the Self.

All activities and obligations that belong to the outer world do definite harm to the secret activities of the unconscious. Through these unconscious ties, those who belong together come together.[238]

The bees forage and work as individuals but they share a bond through the hive and are termed 'eusocial' creatures by research scientists. Modern bee-researchers have noted that one of the hallmarks of the communal life of *Apis mellifera* (the honey bee) is behavioural differentiation among individual worker bees, which results in a division of labour in the hive or colony that serves the life force of the whole and enables all the bees (the multiplicity) to act with 'united force'.[239] They work with altruism for wholeness, that of the hive; we work on our individuation process, and the effects of this have a wider objective healing effect as we saw in the Transformed Berserker section.

During a series of lectures in Switzerland in 1923, Rudolph Steiner acknowledged something central to the mystery of the bees; indeed, perhaps it is the key to understanding why the bees, the honey and the hive have been regarded as so numinous by humankind:

By working mechanically we destroy what Nature has elaborated in such a wonderful way . . . The bee wants always to remain within itself, to stay within the sphere of its own substance. Every external influence is felt as disturbing, as something to be warded off.[240]

His remarks could well be applied to the alchemical *opus* and the individuation process. The bees are intent on circumambulating their own centre and are devoted to the transformative processes of honey-making and to the temenos of the hive with its living golden centre, the egg-

---

238 Von Franz, 'The Process of Individuation', in *Man and His Symbols*, pp. 220f.

239 Gene Robinson, 'Hormonal and Genetic Control of Honeybee Division of Labour', in L. J. Goodman and R. C. Fisher, *The Behaviour and Physiology of Bees*, p. 14.

240 Steiner, *Bees,* p. 3.

laying fertilised queen who, as we shall see here in the *hierosgamos* section, is a hermaphrodite (Figure 26) like the transforming Mercurius who is also the living *lapis* at the end of the alchemical operations.

Figure 26. Hermaphrodite, from the *Rosarium philsophorum* (18th century).

Jung's masterpiece and final work, *Mysterium Coniunctionis,* describes the making of the azure blue quintessence, the *caelum.* The ingredients listed in the 'recipe' for this quintessence include golden flowers (chelidonia or celandine) and other herbs, plus honey as well as blood. This is the mixture in alchemy that fixes the 'heavens below' in the approach to the *unus mundus* and *multiplicatio*—the final stages of the *opus magnum.* I could now focus more creatively on the *caelum* ('the alchemical blue') and the flowers, but instead will stay with this more difficult aspect of circumambulation and the amplification process here in the final stages of this book.

****

Theodor Abt makes the following cogent observations about the 'Secret of the Union' (the *mysterium coniunctionis):*

> *The self is a symbol for the experience of an inner-psychic centre that is in itself light and dark, male and female, a union of the opposites. The final union of soul and the two-one self is again a rebis, a two-oneness, as innumerable symbols of the stone of the sages especially the often mentioned lead-copper confirm. The ego of the adept is then the fourth 'pillar' in this quaternio allowing the* quinta essentia *to come to life or to become real. This would correspond to the divine child or the entirely new attitude. It is a durable connection of the day- and the night-world, of consciousness and the unconscious, of the mortal and the immortal.*[241]

The *Hierosgamos*

The Queen Bee's Nuptial Flight

The queen bee stays in the dark of the hive. She leaves it only two (or possibly three) times in her life. Her initial flights are exploratory, to 'test her wings' and reconnoitre, before going on her nuptial flight when she flies up towards the sun and then, high in the air, mates with selected drones. After she descends and re-establishes herself in the dark protection of her hive, she begins the laying of eggs in the perfectly formed wax hexagonal cells. This egg laying will continue without pause for the

241 Abt, *The Book of Pictures,* pp. 44f.

rest of her life.[242] Maurice Maeterlinck registers the poetry and numinous nature of what is the equivalent of the *hierosgamos* of the queen bee in the following passionate expression.

> *Very few, I imagine, have profaned the secret of the queen-bee's wedding, which comes to pass in the infinite, radiant circles of a beautiful sky. But we are able to witness the hesitating departure of the bride-elect and the murderous return of the bride.*
>
> *However great her impatience, she will yet choose her day and her hour, and linger in the shadow of the portal till a marvellous morning fling open wide the nuptial spaces in the depths of the great azure vault. She loves the moment when drops of dew still moisten the leaves and the flowers, when the last fragrance of dying dawn still wrestles with burning day, like a maiden caught in the arms of a heavy warrior; when through the silence of approaching noon is heard, once and again, a transparent cry that has lingered from sunrise.*
>
> *Then she appears on the threshold—in the midst of indifferent foragers, if she have left sisters [if another queen bee has hatched out] in the hive; or surrounded by a delirious throng of workers, should it be impossible to fill her place [if she is the sole queen].*
>
> *She starts her flight backwards, returns twice or thrice to the alighting-board; and then, having definitely fixed in her mind the exact situation and aspect of the kingdom she has never yet seen from without, she departs like an arrow to the zenith of the blue. She soars to a height, a luminous zone, that other bees attain at no period of their life. Far away, caressing their idleness in the midst of the flowers, the males have beheld the apparition, have breathed the magnetic perfume that spreads from group to group till every apiary near is instinct with it. Immediately crowds collect, and follow her into the sea of gladness, whose limpid boundaries ever recede. She, drunk with her wings, obeying the magnificent law of the race that chooses her lover, and enacts that the strongest alone shall attain her in the solitude of the ether, she rises still; and, for the first time in her life, the blue morning air rushes into her stigmata singing its song, like the blood of heaven, in the myriad tubes of the tracheal sacs, nourished on space, that fill the centre of her body. She rises still. A region must be found unhaunted by birds, that else might profane*

---

242 The fertilised queen is mostly too heavy for flying.

the mystery. She rises still; and already the ill-assorted troop below are dwindling and falling asunder. The feeble, infirm, the aged, unwelcome, ill-fed, who have flown from inactive or impoverished cities, these renounce the pursuit and disappear in the void. Only a small, indefatigable cluster remains, suspended in infinite opal. She summons her wings for one final effort; and now the chosen of incomprehensible forces has reached her, has seized her, and bounding aloft with united impetus, the ascending spiral of their intertwined flight whirls for one second in the hostile madness of love.[243]

A striking aspect of Maeterlinck's description is the way the queen starts her nuptial flight backwards and hesitates as she gets her bearings on leaving the darkness of the hive for the first time. It resonates with how we need to reorient ourselves as we approach a new level or new sunrise in our individuation process. There is so much of the alchemical 'Mystery of the Queen' in this particular celebratory expression of the queen bee's *hierosgamos.*

Renewal of the Life of the Hive

In the life cycle of the hive, swarming and reproduction could be considered to be a symbolic equivalent of what is a called the alchemical 'mystery of the queen'. A bee swarm, where the bees en masse fly out from the hive to found a new colony in a new place (usually in a protected hollow) with a new queen, is a rupture of the bees' daily rhythm. This break is like the end of the *ancien regime* in a revolution. This stage prepares the way for and allows for an absolutely new thing to emerge or break through. This is symbolised in the Chinese Book of Changes, the *I Ching,* in the movement from hexagram 49, 'Revolution' to hexagram 50, 'The Cauldron'[244]—a complete ending followed by a new containment. The bees leave the old hive never to return, but a new colony is then established. This is also what happens in fairy tales when the old king dies. First comes the interregnum, a place and time of flux and flow and creative chaos like when bees swarm. This liminality provides a time and place where the opposing forces can dynamically and creatively

---

243 Maeterlinck, *The Life of the Bee,* pp. 303ff.

244 Richard Wilhelm, *The I Ching or Book of Changes,* pp. 189ff.

come together thorough the agency of the unifying, healing, but ever tricky Mercurius the pontifix. Then comes a renewal, with a new king or queen establishing a new kingdom as shown by the images of the red king in alchemy, for example in the text *Splendor Solis*.[245]

The following observations are also by Maeterlinck:

> *Most creatures have a vague belief that a very precarious hazard, a kind of transparent membrane, divides death from love; and that the profound idea of nature demands that the giver of life should die at the moment of giving. Here this idea, whose memory lingers still over the kisses of man, is realized in its primal simplicity. No sooner has the union been accomplished than the male's [drone's] abdomen opens, the organ detaches itself, dragging with it the mass of the entrails; the wings relax, and, as though struck by lightning, the emptied body turns and turns on itself and sinks into the abyss. . . .*
>
> *. . . She [the queen bee] descends from the azure heights and returns to the hive, trailing behind her, like an oriflamme, the unfolded entrails of her lover . . .*
>
> *. . . and when the breathless queen has reached the alighting-board, some groups will form and accompany her into the hive; where the sun, hero of every festivity in which the bees take part, is entering with little timid steps, and bathing in azure and shadow the waxen walls and curtains of honey. Nor does the new bride, indeed, show more concern than her people, there being not room for many emotions in her narrow, barbarous, practical brain. She has but one thought, which is to rid herself as quickly as possible of the embarrassing souvenirs her consort has left her, whereby her movements are hampered. She seats herself on the threshold, and carefully strips off the useless organs, that are borne far away by the workers; for the male has given her all he possessed, and much more than she requires. She retains only, in her spermatheca, the seminal liquid where millions of germs are floating, which, until her last day, will issue one by one, as the eggs pass by, and in the obscurity of her body accomplish the mysterious union of the male and female element, whence the worker-bees are born. Through a curious inversion, it is she who furnishes*

---

245 I would like to acknowledge Andreas Schweizer's expression of this significance in lectures on 'The Teaching of Komarios and Cleopatra: Reflections on a Greek alchemical text dedicated to Komarios and Cleopatra'.

the male principle, and the drone who provides the female. Two days after the union she lays her first eggs, and her people immediately surround her with the most particular care. From that moment, possessed of a dual sex, having within her an inexhaustible male, she begins her veritable life; she will never again leave the hive, unless to accompany a swarm; and her fecundity will cease only at the approach of death.[246]

Maeterlinck is highly poetic in his expression of awe for the unfolding of some of the self-regulatory processes at work in the beehive. He describes in detail how the bees dedicate their energy to what he calls the 'future of the hive' and to what Karl Kerényi calls 'the indestructibility of life'. This imagery of the hermaphrodite and the duplex nature of the fertilised queen have much to do with the processes of Mercurius and the fertilising union of solar and lunar in what is referred to as the alchemical 'Mystery of the Queen'. The image of the hermaphroditic fertilised queen bee is particularly evocative of Mercurius.

There is a death or dying motif of the *hierosgamos* that belongs psychologically to a world of cyclical regeneration in which the Great Goddess or Mother Goddess and the feminine principle of eros are predominant. Such a world is described by the symbolism of the bee, the honey and the hive in the predominance of the queen and the worker bees. In patriarchal warrior cultures, this motif of death does not have such prominence.[247] In the *hierosgamos* of the queen bee, the drone dies a dramatic, dismembering death in mating with the queen.

The Alchemical Hierosgamos, *the Divine Water and the Gold*

The consummation of the alchemical *hierosgamos* is the 'earthing' of the spirit and the spiritualising of the earth, or as Jung writes,

the union of opposites and reconciliation of the divided, in a word, the longed-for act of redemption whereby the sinfulness of existence, the original dissociation will be annulled in God.[248]

Such widening of consciousness initially involves upheaval and dark-

---

246 Maeterlinck, *The Life of the Bee,* pp. 306, 316, 318ff.

247 Von Franz, *On Dreams and Death,* p. 55.

248 *Mysterium Coniunctionis,* CW 14, par. 207.

ness, but then, as Jung says, there is 'a broadening out of man to the whole man. This "Man", being indescribable, is an intuitive or "mystical" experience, and the name "Anthropos" is therefore very apt because it demonstrates the continuity of this idea over the millennia.'[249] We have seen an example of this in the Transformed Berserker vision of Niklaus von Flüe.

Von Franz says that the *hierosgamos,* the sacred marriage, is also a death wedding (because, for a new unity to be born, something old must die at the same time, meaning that the *rubedo* and the *nigredo* are synchronous or uroboric). She notes that this is a widespread archetypal motif, that of the union of the upper and lower in the form of bridegroom and bride.[250] It represents the union of the psychic opposites and, as in the following extract from the Komarios text—of the upper and the lower, or the earth and the sky.

> *12. Look, wise ones, and understand; observe the fulfilment of the art through the bringing together and union of the bridegroom and bride. Observe the plants and their differences. I say to you: see and understand that out of the sea the clouds arise, and the clouds carry the blessed waters with which they water the earth and from the earth spring forth seeds and flowers. Thus our cloud, which rises from our element, also carries divine water and waters the plants and elements and needs nothing that comes from other places.*
>
> *13. Observe the paradoxical mystery, the great unknown; observe how obvious the truth is to you. Observe and water your earth and be careful how you nourish your seeds, so you will be able to harvest the most beautiful fruit . . . [then the process of transformation is described] and when the soul (psyche) and spirit (pneuma) have become one, then project it (the one) onto the body of the silver and you will have gold, such as kings in their treasure houses do not possess.*[251]

Von Franz also points out that the alchemists praise the *lapis* in their writings as a 'living thing that streams forth "rose-coloured blood" or

---

249 *Mysterium Coniunctionis*, CW 14, pars. 207ff.

250 Von Franz, *On Dreams and Death*, p. 42.

251 'Komarios to Cleopatra', ibid.

"rose colour" and has a healing effect on its environment', like the honey overflowing from the Transformed Berserker—the unified figure, in which the divine opposites have come spontaneously together and so could have a healing effect in the world. 'This is a love, a deep eros or connectedness, that flows from the individual human being who has become whole.'[252]

Out of the mysterious union, the *hierosgamos*, as Gotthilf Isler has said, comes the birth of a new being, the new king or a divine child, the *filius philosophorum* (son of the sages or philosophers),[253] and all these births are various aspects of the Self, as too is the *lapis*—the gold that the alchemists longed and worked for, the all-healing tonic, the never-ending victuals, the light which cannot be extinguished. To me this is what the honey elixir represents. It is an eternal unalloyed substance, which when used in the right measure preserves, purifies, nourishes and heals. Jung wrote:

> *Philosophic gold is an embodiment of psyche and pneuma, both of which signify "life spirit". It is in fact an* aurum non vulgi, *a living gold, so to speak, which corresponds in every respect to the* lapis. *It too is a living being with a body, soul and spirit and it is easily personified as a divine being or superior person like a king [or queen], who in olden times was considered to be God incarnate.*[254]

This is very much connected with the description of the honey or balsam—repeated here as a lyrical refrain, the ground water of this book:

> *In the honey, the 'sweetness of earths', we can easily recognise the balsam of life that permeates all living, budding, and growing things. It expresses, psychologically, the joy of life and the life urge which overcome and eliminate everything dark and inhibiting. Where spring-like joy and expectation reign, spirit can embrace nature and nature, spirit.*[255]

---

252 Von Franz, 'The Transformed Berserker', in *Archetypal Dimensions of the Psyche,* p. 49.

253 Isler, G. 'The Snake as a Spirit of the Earth', lecture.

254 *Mysterium Coniunctionis,* CW 14, par. 354.

255 Ibid., par. 698.

Deep Eros and the 'Mystery of the Queen'

As with the mystery of the life force in a beehive, the psychological union of conscious and unconscious symbolised in alchemy by the royal marriage must remain a mystery. Jung sums this up at the end of the 'Rex and Regina' section in *Mysterium Coniunctionis,* saying that the 'Mystery of the Queen' is the secret of the art:

> *The art is queen of the alchemist's heart, she is at once his mother, his daughter, and his beloved, and in his art and its allegories the drama of his own soul, his individuation process, is played out.*[256]

There is an eros too deep for words underlying the *hierosgamos.* The mystery of the great love between the Queen of Sheba and King Solomon is also beyond words, but something of its joy and essence is expressed in the 'Song of Songs'—a lyrical text, which some alchemists understood (as in *Aurora Consurgens)* as expressing the feeling of great jubilation at the completion of their *opus.* According to Marie-Louise von Franz: 'It is a description of the completed individuation process, of an ultimate union of psychic opposites, a liberation from all egocentricity and an ecstatic entrance into a state of divine wholeness.'[257]

Penotus in the following description reveals the significant presence of Mercurius who, Jung says,[258] as *aqua permanens* or spiritual water is a life principle and the 'marriage maker' between man and woman in this union and this 'love strong as death':[259]

---

256 Ibid., pars. 542f. In pars. 533f., Jung says that the queen is what surrounds the sun. The relationship of great eros between the Queen of Sheba and King Solomon is 'the prototype of spiritual *Minne',* Johannes Grasseus says , "This is the chaste, wise, and rich Queen of Sheba, veiled in white, who was willing to give herself to none but King Solomon. No human heart can sufficiently investigate all this'. Jung refers to the queen appearing in the alchemical texts as 'the maternal vessel of Sol'. He goes on to say that: 'As a matter of fact the queen corresponds to the soul *(anima)* and the king to spirit, the dominant of consciousness.' In view of the alchemists' interpretation of the queen, he adds, 'we can understand why the secret of the work was sometimes called the "Reginae Mysteria".'

257 Von Franz, *On Dreams and Death,* p. 45.

258 *Mysterium Coniunctionis,* par. 658.

259 Song of Songs, 8:6.

You have the virgin earth, give her a husband who is fitting for her! She is the Queen of Sheba, hence there is need of a king crowned with a diadem—where shall we find him? We see how the heavenly sun gives off his splendour to all other bodies, and the earthly or mineral sun will do likewise, when he is set in his own heaven, which is named the "Queen of Sheba", who came from the ends of the earth to behold the glory of Solomon. So, too, our Mercury has left his own lands and clothed himself in the fairest garment of white, and has given himself to Solomon, and not to any other who is a stranger [extraneo] and impure.[260]

Consider also the imagery of "the fairest garment of white" (a solification garment) and of the bride in the following numinous dream a contemporary woman had in the first stages of analyst training:

I had a sense there was to be a wedding. I walked out of the room and spoke to Regina and she suggested I could go and try on my bridesmaid's dress. She was doing housework preparing for a special meal, a celebration. I went into a bathroom and saw a voluminous white dress, as it was here in readiness for tomorrow. I asked did I have to try it on and she said, 'Yes, I think you will like doing this.'

I went to look in the room where the dresses were hanging protected with covers and my dress was large and white and had many pearls sewn on it. It was the bride's dress [rather than a bridesmaid's dress]. A. [the analyst] had been called away but I knew he would return and then we could resume our work together. We had a great amount of affection for each other. I looked at drawings, the designer's sketches, of the dresses in plastic sleeves in a folder. They were photocopies so I wouldn't have to handle the actual garments. Then I went into a special dressing room and saw the real dress. It was beautiful. At first I was worried it might be too big for me, but I trusted the dressmaker would be there today to make it fit me perfectly and well, so I relaxed. I felt very sure of my feelings for A. and the connection, the relationship, between us was the main thing. I knew he would return before the wedding.

This dream imagery is a sustaining example of the living mystery and the secret of the work, which was sometimes called in alchemy the *'Reginae Mysteria'* (the 'Mystery of the Queen').[261]

---

260 *Mysterium Coniunctionis,* CW 14, par. 533.

261 Ibid., par. 536 and n.429.

Multiplicatio and Flowering

The symbolism of flowering is important in the final stages of the alchemical *opus magnum,* as it is in nature. In nature, the plant, for example a fruit tree, flowers and then the bees pollinate it. They fly to the centre (usually golden) of the blossom, hum busily, and 'work' the flowering plant—and then the fruit sets. As we well know, this activity is focused on the centre and serves the queen who lays the eggs and thus renews the life force of the hive. The result of the alchemical *opus* is often called its *fruit*.[262] In love there are sufferings to bear as well as joys. Through passionate love and the experience of deep eros, one moves beyond matters of the ego and this can lead to the birth of the Self.[263] Miguel Serrano recalls Jung musing, as if to himself, of mystic weddings,[264] a process of mutual individuation:

> *Somewhere there was once a flower, a Stone, a Crystal, a Queen, a King, a Palace, a Lover and his Beloved, and this was long ago. On an island somewhere in the ocean five thousand years ago . . . Such is Love, the Mystic Flower of the Soul. This is the Centre, the Self.*[265]

Rudolph Steiner's comment on the connection of bees with flowers provides some further amplification of the eros theme in connection with the bees and the flowers and sexuality:

> *They [the bees] have indeed carried into the hive that which lives in the flowers. When you begin to think through all of this properly, you will have unlocked the whole secret of the beehive. The living element of this thriving, germinating love that is spread out over the flowers is also contained in the honey that the bees make.*[266]

The queen bee as the living centre of the hive gives the parts of the whole a purpose and a completeness. Thus, not only the life force of the community of bees is ensured, but also that of the whole ecosystem—the

---

262 Ibid., par. 181, n. 321, *The Turba.*

263 Von Franz, *Hall Ar-Rumuz (Clearing of the Enigmas),* p. 48.

264 Claire Dunne, *Carl Jung: Wounded Healer of the Soul,* pp. 171f.

265 Serrano, *C.G. Jung and Hermann Hesse: A Record of Two Friendships,* pp. 60f.

266 Steiner, *Bees*, p. 3.

regeneration, fertility and *multiplicatio* linked with the cyclical flow (a *circulatio*) of the Mother Goddess's feminine potency. It is the dark fertile earth from where the birth of new light and eros springs; or arrow-like, shoots up or sprouts—ascending and moving forward and up, reaching for the sky (like the queen bee on her nuptial flight) and the sunlight (the solar consciousness). Consciousness is fertilised by the waters of the unconscious and its centre, the Self.

To further consider the imagery of *mutiplicatio,* a helpful example is that of corn: a plant connected symbolically with the mystery of life and death, and of transformation or regeneration. In the Bible Jesus speaks of the grain of corn falling into the ground and dying there, but then bringing forth much fruit (John 22:24).[267] The analogy in alchemical writing is referred to as the problem of *multiplicatio.* Von Franz elaborates on this:

> *The alchemists say that at a certain moment when the philosopher's stone is made, the vessel is opened and the stone begins to emanate a transforming activity, and every metal touched by it is transformed into gold. The analogy is that the philosopher's stone is made inside a vessel, where it goes through dissolution in the darkness and then resurrects. The vessel [for example, the skin of a beast in the bugonia] is opened when the stone develops, transforming into a new radiating and emanating in what the alchemists called the process of multiplication. The psychological analogy seems to have to do with the fact that when one succeeds consciously and positively to relate to an archetypal constellation, the effect is widespread.*

For example, if the rainmaker or the medicine man deeply introverts and gets in touch in the right way with the powers of the Beyond, even such things as the needed rain falling over the whole country can occur. Confucius said that if the noble man sits in his room and has the right thoughts and writes down the right things, he is heard a thousand miles around.[268] In Samson's story, the sweetness, refinement of the golden honey, came out of the drives and the physical strength. We need the fire and the passion and the ego strength to structure these raw materials, as

---

[267] Von Franz, *The Feminine in Fairy Tales,* p. 159.
[268] Ibid.

well as needing the sweetness. These things are the *prima materia* of the alchemical transformation process. We must experience life and our own emotions and wildness (that von Franz says every saint, when speaking of the Transformed Berserker vision of Brother Klaus, would experience in the extreme), before we can reach the *lapis* or honey stage.

Ben Jonson wrote:

> *Beside, who doth not see, in daily practice,*
> *Art can beget bees, hornets, beetles, wasps,*
> *Out of the carcasss and dung of creatures,*
> *Yea, scorpions of an herb, being rightly plac'd.*[269]

I have already discussed some aspects of another form of *multiplicatio*—bugonia, the spontaneous regeneration of bull-born bees. In ancient Rome the practice of *apes facere* (making bees) was one process through which the creation could be re-enacted by man. Creating bees in this way was another version of the *multiplicatio* symbolism,[270] best performed when the sun enters Taurus, that is, in springtime. In astrology, Taurus, ruled by the planet Venus, is the sign of the bull and, as I've mentioned, this indicates fixed earth (an enduring manifestation on the material level). The glyph for Venus is made up of the circle of the sun with the crescent moon resting on its head—an image of the *coniunctio;* while for Mercury, the glyph is the crescent moon on the sun, on the earth—very apt symbolic imagery for the role of Mercurius in the *hierosgamos.* This earthing is the key element is the shaping of the *caleum* (that is, the making of 'the heavens below' and the *unus mundus*). As a professional astrologer and Jungian analyst, and as a human being and a pilgrim, it is important for me to remember that the earth works together with the sun and moon in the transformation and healing that flows from Mercurius.

****

The psychological implications and images of the multiplication stage are most interesting as transformative effects emanate, induction-style,

---

269 Jonson, *The Alchemist*, II, iii.

270 See also Kerényi's 'Light and Honey' chapter in *Dionysos.*

from the activated Self in a process of conscious realisation.[271] Compare this with the reality and the symbolism of the effect of the queen bee's presence in the hive of bees:

> *The queen, for her part, is the unifying force of the community; if she is removed from the hive, the workers very quickly sense her absence. After a few hours, or even less, they show unmistakable signs of queenlessness.*[272]

Psychologically speaking, events, no matter how ordinary or humble, take on an importance when they participate in the process of individuation. It is like taking small steps, daily steps, on a pilgrimage. The sacred centre is reached eventually. *Multiplicatio* can shed some light on how psychoanalysis or psychotherapy might do its healing work, but this essentially also remains a deep mystery. To some extent, the consciousness of an individual (for example, the analyst who is related to the Self and pays daily attention to the symbolic life and also is 'down to earth') seems to be contagious and tends to multiply itself in others (for instance in analysands and friends). Jung states:

> *We seldom find a high point of religious feeling where this eternal image [of the extreme opposites] of the royal marriage does not appear . . . what the union of opposites really 'means' transcends human imagination.*
>
> *. . . whenever this eternal image or archetype is obscured, a person's life loses its proper meaning and consequently its balance. So long as he knows that he is the carrier of life and that it is therefore important for him to live, then the mystery of his soul lives also—no matter whether he is conscious of it or not.*[273]

Here, once again, we have the celebration of the flux and flow of the life force—*zoë*. The individual bees make a beeline to the flowers, that is, they fly in the most direct straight lines; but they also dance mandalas and lemniscates to communicate to the hive. As Rudolf Steiner notes:

> *The bees have 'indeed carried into the hive that which lives in the flowers. [And when you begin to realise this, you begin to unlock the] whole secret*

---

[271] Edinger, *Anatomy of the Psyche*, p. 227.

[272] Newman, *Man and Insects,* epigraph, chap. 1.

[273] *Mysterium Coniunctionis,* CW 14, pars. 200f.

of the beehive. The living element of this thriving, germinating love that is spread out over the flowers is also contained in the honey.[274]

Marie-Louise von Franz says of the wise Cleopatra in the Komarios text that in this female alchemist a great secret is hidden.[275] She writes that it was the secret of creation and the alchemical secret of repeating the creation which was hidden within the feminine figure, in this case the wise Cleopatra. In the Komarios text, the sage (alchemist) Ostanes and his companions address the following words to Cleopatra. They are asking that she, through her feminine wisdom and knowledge, illuminate for them the world of matter and the flux of life and creation:

In you [Cleopatra] is hidden the terrific and wonderful secret. Make known to us how the highest descends to the lowest, and the lowest ascends to the highest, and the midmost draws near to the lowest and the highest, so that they are made one with it; how the blessed waters come down from above to awaken the dead, who lie round about in the midst of Hades, chained in the darkness; how the elixir of life comes to them rousing them out of their sleep.[276]

Cleopatra answers:

When the waters come in, they awaken the bodies and the spirits, which are imprisoned and powerless. . . . Gradually they bestir themselves, rise up, and clothe themselves in bright colours, glorious as the flowers in spring. The spring is glad and rejoices in the blossoming ripeness they have put on.[277]

'When the waters come in' is indeed a time for exaltation and celebration. I had the following dream towards the end of writing this work:

Fresh clear blue water was flooding up out of a centre point: an underground spring and flowing out over my land. It was even flowing upstream in a dry river bed or channel, moistening the land that had been parched by drought.[278]

---

274 Steiner, *Bees,* p. 3.

275 Von Franz, *Creation Myths*, p. 228.

276 Ibid., pp. 227f.

277 Ibid., p. 228.

278 I had this dream while attending a two-day seminar on 'The subjective in Jungian psychology: a dialogue between analysts and academics' in Australia. On an objective

Centring: Circumambulating the Centre

Mandala Symbolism: Rotation and Circulatio

The bees have a dance language; they move in circles and lemniscates to communicate where the nectar is. Their foraging flights go out from and return to a centre—the hive. A bee retinue circumambulates the queen at all times. The focus is on movement around the numinous, life-giving and sustaining centre. This brings us to the mandala and to the archetype of inner order in depth psychology.

A mandala is a magic circle, symbolising psychic totality and expressing the pattern of life around the centre. Mandalas exist all around the world, often in a sacred or ritual context, and are well known in the colourful and intricate sand mandalas of Tibetan Buddhism and the rose windows in Christian cathedrals such as the impressive one at Chartres. In India mandalas were used as yantras, aids to meditation. Their structure is usually a fourfold one within the quarter circles. They often take a form based on an open flower, the cross or the wheel. The deity is at the centre and is the focus of a sacred circumambulation.

In his essay 'Concerning Mandala Symbolism', Jung writes of the mandala:

> *There are innumerable variants of the motif . . . but they are all based on the squaring of a circle. Their basic motive is the premonition of a centre of personality, a kind of central point within the psyche, to which everything is related, by which everything is arranged, and which is itself a source of energy. The energy of the central point is manifested in the almost irresistible compulsion and urge to become what one is, just as every organism is driven to assume the form that is characteristic of its nature, no matter what the circumstances. This centre is not felt or thought of as the ego but, if one may so express it, as the self.*[279]

---

level much of Australia was drought-stricken at this time and there were severe water restrictions, but in a community of colleagues there is always a question of whether the 'water of life' flows, and of how we work today (as Jungian therapists, healers or educators) with the symbolic life or the archetypal images that come up from the depths to consciousness through dreams, for example.

279 *The Archetypes and the Collective Unconscious,* CW 9i, par. 634.

Mandalas are not only painted or drawn or made out of coloured sand as the monks in Tibet and the Navajo Indians do. Mandalas can also be danced as the bees do. In his 'Commentary on *The Secret of the Golden Flower',* Jung wrote:

> *Among my patients I have come across cases of a woman who did not draw mandalas but danced them instead. In India there is a special name for this:* mandala nrithya, *the mandala dance. The dance figures express the same meanings as the drawings. My patients can say very little about the meaning of the symbols but are fascinated by them and find that they somehow express and have an effect on their subjective psychic state.*[280]

Psychologically, rotation or *circulatio* is a manifestation of the energy of the central point within the psyche, which has in itself a kind of energy or is as Jung describes it a 'source of energy'. As Regine Schweizer said in a lecture, that energy has an effect on the outside, on the ego. It manifests in a strong compulsion and urge to become 'what one is'.[281]

On the other hand, this rotation or circulation leads to or means a kind of development in the unconscious too. In general the motif means development, which takes place in the background of the psyche. We can also say that rotation or circulation means creation. 'The new light, the new life or inner value, wants to be born and now wants to incarnate in a man's or a woman's psyche'.[282]

In the magic papyri, Hermes is invoked as follows: "Oh, Hermes, ruler of the world, thou who dwellest in the heart, circle of the moon, round and square." In alchemy, Mercurius is the rotundum par excellence.[283]

Astrologically, circulation refers to the movement of the repeated ascent and descent of the planetary spheres whereby each of the archetypal principles, symbolised by the planetary archons of the twelve signs of the zodiac, is encountered successively. Chemically, *circulatio* is the process in which a substance is heated in a reflux flask. The vapours ascend and condense: then the condensed fluid is fed back into the belly of the flask,

---

280 *Alchemical Studies,* CW 13, par. 32.

281 R. Schweizer, 'The Divine Girl', lecture, St Niklausen, Switzerland, 2005.

282 Ibid.

283 *Mysterium Coniunctionis,* CW 14, pars. 166f.

where the cycle is repeated.

Edinger describes circulation psychologically as the repeated circuit of all aspects of one's being, which gradually generates awareness of a transpersonal centre uniting the conflicting factors [the warring opposites].[284] There is a transit through the opposites, which are experienced alternately again and again, leading finally to their reconciliation that Jung describes thus: 'The painful suspension between opposites gradually changes into the bilateral activity of the point in the centre'.[285]

****

The mysteries of the queen, and those around the life of the bees and the hive, remain living mysteries. And humankind will no doubt 'continue to search for truths about themselves in the bees and the gold of the honeycomb'.[286]

Marie-Louise von Franz writes:

> *When the ego is not in union with the deeper layers of the instinctive personality, ...it gets torn between the opposites. If the ego could relate directly to the Self, which is a unifying symbol [like the Queen Bee], the conflict would fade and the ego would again function in wholeness. That is the normal way in which the opposites function, and the main impulse is again the flow of life which comes from the totality. A conflict is never really solved, the emotion invested in it diminishes, one outgrows it by suffering and it becomes absorbed in a new form of life, with the result that one looks back dispassionately on it from a different angle.*[287]

The *rubedo* stage of alchemical transformation (that in which the red–gold is produced) is a stage of coagulation or completion. It is also linked with descent, earth and circulation.[288] The king or the *lapis* as bridegroom of the white (feminine) bride is red in colour. This new life of the *rubedo* (redness) flows from the Self not the ego. New life and emotion

---

284 Edinger, *Anatomy of the Psyche*, pp. 142f.

285 *Mysterium Coniunctionis,* CW 14, par. 296.

286 Wilson, *The Hive,* p. 271.

287 Von Franz, *Shadow and Evil in Fairy Tales,* p. 38.

288 Edinger, *Anatomy of the Psyche,* p. 117.

have begun to flow again after the darkness of the *nigredo*.[289] The *rubedo* and the *nigredo* exist almost together, as Andreas Schweizer says:

With mere effort of will, no persisting transformation will take place . . . There is no causal relationship between putrefaction and conception. Any relation between them would instead be a 'synchronistic' one. The death of the old is at the same time the birth of the new; the conception of the new conjoins the death of the old. This then means that rather than a transformation process as such, it could be seen more like a spiral rotation out of a nameless centre through which a process of creation takes place.[290]

As with the legendary *bugonia*—spontaneous new life of bees regenerated from the carcass of a bull, ox or lion—the idea is that alchemy recreates the creation. The rotation around a centre is once again the dynamic of a new life and a new, green sprout coming from a seed buried under the earth.

In a lecture to analysts-in-training, Dr Schweizer added:

This means that you cannot become an alchemist or an analyst without committing yourself to the process as fully as possible, without being personally involved in the work with your entire life and personality. Psychologically, the alchemists' work in the laboratory resembles the analytical work. Theories may help to understand certain things, but they are not really important. Only if you carefully consider and honour your own, very personal experiences will a process of transformation start. In this regard the matter of the alchemist, his prima materia, is the factual life in its 'just-so-ness', beyond the illusions that we all have about ourselves.

As the alchemist finds his gold or stone on the dunghill or in the ashes—that is, as he discovers the concealed creative activity of God in everything that exists—so the naked truth of our life is the often-painful starting point of the most important changes in our life.[291]

---

289 Andreas Schweizer, 'The Teaching of Komarios and Cleopatra', lecture.

290 Ibid.

291 Ibid.

Dorn: The Caelum *and Three Stages of Illumination or* Coniunctio

Jung mentions three stages of union and illumination in *Mysterium Coniunctionis* that compare with the three stages of the *coniunctio* in Gerhard Dorn's alchemical writings. It is important to remember that Mercurius is not only the *lapis* as *prima materia,* but also the *lapis* as *ultima materia*—the goal of the *opus.* Albertus Magnus (citing Gerber) says: 'One is the stone, one the medicine, and therein lies the whole magistery'. Of this, Jung commented: 'In these words Albertus Magnus, more than 300 years ahead of Dorn, describes the celestial substance—the balsam[292] [the honey] of life, and the hidden truth'.[293]

This description of Dorn's has even earlier roots in Greek alchemy (for example, in the work of Komarios) but describes a transcendent substance characterised by a large number of opposites, or as Jung says, 'antinomies'.[294] This is the case, Jung says, because unequivocal or definitive statements can only be made regarding immanent objects. On the other hand, transcendental ones can only be expressed by paradox. The fact that they are duplex needs to be honoured: they are and they are not—they are both, and.

In the individuation process there comes a stage where an individual gains trust in the ability of the Self to sustain one. The individual (or, in analysis, the analysand) is then capable of self-reliance, having attained what the alchemists call the *unio mentalis.* This is often represented by a mandala image that alludes to the sky and the stars and to something like an 'inner heaven' (firmament or microcosm). Jung writes of the mandala:

> *This, too, is that circular product, the* caelum, *which Dorn wanted to produce by 'assiduous rotary movements'. To shape out or twist out 'his intuition of a mysterious centre pre-existent in man, which at the same time represented a cosmos', i.e. a totality, while he himself remained conscious*

---

292 Compare Jung's passage: 'In the honey, the "sweetness of earths", we can easily recognise the balsam of life that permeates all living, budding, and growing things. It expresses, psychologically, the joy of life and the life urge which overcome and eliminate everything dark and inhibiting. Where spring-like joy and expectation reign, spirit can embrace nature and nature, spirit.' *Mysterium Coniunctionis,* CW 14, par. 698.

293 Ibid., par. 714.

294 Ibid., par. 715.

that he was portraying the self in matter. He completed the admixture of honey, magic herbs and human blood, or their meaningful equivalents. Just as a modern man does when he associates numerous symbolic attributes with his drawing of a mandala.[295]

Based on the correspondence between plants growing on the earth, and the stars and planets in the sky and their astrological signatures, Dorn evolved what von Franz calls 'an alchemical piece of active imagination, obtained by mixing substances, which is a symbolism of the process of individuation'. This refers to the *caelum,* and 'shaping the heavens below', which Jung says prefigures the appearance of the Self—the unity at the end of the work. In this shaping there are clear allusions to the sky and stars, and thus to the inner heaven—the microcosm. (And meaningful equivalents, the influence of the heavens as well as the underworld, are drawn into Dorn's quintessence.)

Honey, golden flowers (chelidonia or celandine), rosemary flowers, the Mercurialis plant, the red lily and human blood are added to the mixture and in this way, writes von Franz,

> *the heavens below are established. This mixture must be cooked [in much the same way as the bees cure the honey] and the result is the last stage in making the philosopher's stone [the* lapis*]—namely the ultimate union with the whole cosmic nature. When the heavens below have been produced, one is united with the* unus mundus, *cosmic divine nature.*[296]

So we see the mandala and *caelum,* and the *lapis* and the honey, are linked through a circular movement of dance or circumambulation around the centre as the highest value. Thus alchemy meets the bee and the *quadratura circuli,* the square in the circle or the circle in the square—the heaven (circle) meets the earth (square). Consider Jung's words when asked in an interview to make some observations on the term mandala:

> *The mandala is just one typical archetypal form. It is what they call in alchemy the* quadratura circuli, *the square in the circle or the circle in the*

---

295 Ibid., par. 757. Dorn: *caelum* as a mysterious centre: a blue centre as in my dream (see following, at the end of this section).

296 Von Franz, M-L. *The Feminine in Fairytales*, pp. 129–30.

square. It is an age-old symbol that goes right back to the prehistory of man. It is found all over the earth and it expresses either the deity or the self. These two terms are psychologically very much related—which doesn't mean that God is the self or that the self is God. I simply state that there is a psychological relation between them. There is plenty of evidence for this.

The mandala is a very important archetype. It is the archetype of inner order, and it is always used in that sense, either to make an arrangement of the many, many aspects of the universe, a world scheme, or to make a scheme of our psyche. It expresses the fact that there is a centre and a periphery, and it tries to embrace the whole. It is the symbol of wholeness. So you see, when during the treatment there is a great disorder and chaos in a man's mind, this symbol can appear in the form of a mandala in a dream, or else he makes imaginary, fantastical drawings or something of that sort. The mandala appears spontaneously as a compensatory archetype, bringing order, showing the possibility of order. It denotes a centre which is not coincident with the ego but with the wholeness I call the self—this is the term for wholeness. I am not whole in my ego, my ego is a fragment of my personality. The centre of a mandala is not the ego, it is the whole personality, the centre of the whole personality. The mandala plays a very great role in the East, but in our Middle Ages equally. Then it got lost, and was thought of merely as a sort of allegorical, decorative motif. But as a matter of a fact it is a highly important and highly autonomous symbol that appears in dreams and so on, or in folklore. We could easily say that it is the main archetype.[297]

Psychologically this process represents a further stage in the integration of the Self, that is, in the individuation process. Previously the main thing was the union of the soul and the spirit, which was the first stage of the conjunction in which the soul and the spirit are unified and are separated from the body (the *unio mentalis).* A relationship is developed between the ego and the unconscious and, as Jung says, the goal of this stage is 'the attainment of full knowledge of the heights and depths of one's own character'.[298] But then a new problem is constellated.

This involves the freeing of the body itself. This is the stage when the

297 'The Houston Films', in *C. G. Jung Speaking*, pp. 327f.

298 *Mysterium Coniunctionis,* CW 14, par. 674.

introversion or separation of the first stage can move to a new incarnation or reunion of the *unio mentalis* with the body. This is the *unio corporialis* stage in which the insights gained in the first stage can be made real. In Jung's words, 'the second stage of conjunction therefore consists in making a reality of the man who has acquired some knowledge of his paradoxical wholeness'.[299] According to the alchemist Dorn, this is when the body must be distilled until it turns into a blue liquid called heaven *(caelum).* The parallel action to this in *Aurora Consurgens* is the volatilisation or sublimation of the body, which is accomplished by the bride giving it wings and clothing it in royal garments, thus resolving the opposition between 'body' and 'spirit'. This is linked with resurrection.

****

The bees' work in the warm months involves the symbolism of ascent and descent in their daily task of flying up and out of the hive to the flowers and then returning to the hive—descending to go into the entrance usually at the bottom of the hive. They are also focused on the centre and do not stray from this greater purpose and innate guidance.

****

Then there is a third stage of union or conjunction. To begin to understand what this might be, consider the creation of the world. Consider the idea that, in much the same way as a plan exists in the architect's mind before building begins, so God, as supreme and divine architect, conceived a mental image of the world and of everything in it, from which later developed the multiple realisations in matter. This sum of detailed concepts—'all still in one' in this world—is identified in Scholastic teaching with the feminine archetypal figure of the Wisdom of God (Sophia), and was called the sum of the *typi,* or *archetypi ideae.* Dorn calls this conceptual divine world the *unus mundus,* the one cosmos. By the *unus mundus* is expressed the idea that a multiplicity of objects only come into existence by the realisation of such a plan or blueprint.

Before this model plan of the world was materialised, there were no single objects. Instead it was as if germs or seeds of potentiality were

---

299 Ibid., par. 679.

united in a mother liquid, or something similar, rather than there being an immense sum of material objects as in the material world. Thus the 'oneness' of the Godhead in matter was also expressed by the idea of the *unus mundus*. This one world, hidden behind our material world (Figure 27, next page) is considered to be the cosmos with which the alchemist unites himself and becomes one with, but not in a pantheistic form. Pantheism is the idea of becoming one with the actual physical world, but Dorn's idea is rather to unite with what is behind this physical world—the germ of oneness behind the multiplicity of actual existence. Such a 'oneness' is what the bees express in their unity in multiplicity (through 'the wisdom of the hive'). Jung wrote:

> *Although the alchemist thought he knew better than anyone else that, at the Creation, at least a little bit of the divinity, the* anima mundi, *entered into material things and was caught there, he nevertheless believed in the possibility of a one-sided spiritualisation, without considering that the precondition for this is a materialisation of the spirit in the form of the blue quintessence. In reality his labours elevated the body into proximity with the spirit, while at the same time drawing the spirit down into matter. By sublimating [raising] matter, he concretised spirit [brought it down to earth].*[300]

The union of above and below, 'where spirit can embrace nature and nature, spirit', has been an important theme in this investigation into the symbolism of the bee, the honey and the hive. As I started, I will end this work of circumambulation with Jung's jubilant recognition of the importance of the honey:

> *In the honey, the 'sweetness of earths', we can easily recognise the balsam of life that permeates all living, budding, and growing things. It expresses, psychologically, the joy of life and the life urge which overcome and eliminate everything dark and inhibiting. Where spring-like joy and expectation reign, spirit can embrace nature and nature, spirit.*[301]

****

300 Ibid., par. 764.

301 Ibid., par. 698.

Following is the last part of a long dream related to my own training analysis and individuation process. At the end of the dream I went with my analyst to a room where much creative work was in process. This was the fourth room visited in the dream. The first was an intimate space where I sat with my analyst at a round white metal table with a marble or alabaster stone top. The second was a fruit shop (open on two sides) with three Italian-speaking women, and (the dreamer) was the fourth woman.

The third room (also open on two sides) was a richly decorated ornate royal room like in the palace of Versailles, with rich brocade and golden ornaments and some young girls playing. Before entering the fourth room or fifth place, I waited in a grey, hospital-style corridor until my analyst returned from buying bread and then we entered the fourth room and the fifth (quintessential) place together. The fourth room was a bigger, private room in between the first and the third rooms.

Figure 27. Alchemical image of a spiritual pilgrim breaking through the veil of the material world to glimpse the transcendent (17th-century woodcut). It is an image of the *unus mundus.*

The main item of the three things in the room was a big painting [a mural about six metres by three metres] with a bay [at the ocean's edge] showing just blue water merging with blue sky with a few rocks on the shore to the bottom right, and in the top left, a bird like a blue heron. Underneath or behind the main board was a list of personal goals also written in handwriting on a large moveable white board (these were hidden but private goals). This showed me just how much energy and creativity goes into 'the preparation'. The blue heron was the key to this current work. The centre part of the big panel was empty but was an azure blue colour (the sky and sea were indistinguishable). I had a sense that the central section of this artwork would be completed (and the space would be filled in) by me as my work progressed.

This particular dream evokes for me the creation of the *caelum*—the alchemical blue or quintessence. This stage, according to Edinger, indicates when 'one's own unique subjectivity and its supreme value is rediscovered'.[302] It reminds me of what Jung wrote about blue:

There was a blue sky, like the sea, covered not by clouds but by flat brown clods of earth. It looked as if the clods were breaking apart and the blue water of the sea were becoming visible between them. But the water was the blue sky. Suddenly there appeared from the right a winged being [Philemon] sailing across the sky . . . He had the wings of the kingfisher with its characteristic colours.[303]

Its amplification also evokes those ingredients to be mixed with the *caleum* or 'new heaven', which Jung writes of in *Mysterium Coniunctionis,* including honey as well as the golden chelidonia or celandine flowers.

My dream became an inspiration for the preparing, shaping and colouring of this work on the bees, themselves evoked through the golden centres of the flowers, as well as through the Golden Flower, a symbol for the Self in Chinese alchemy.[304] This leads us to the next alchemical stage—the yellow or golden stage.

---

302 Edinger, *The Mysterium Lectures*, p. 288.

303 *Memories, Dreams, Reflections,* p. 176.

304 Edinger, *The Mysterium Lectures*, p. 289.

Eros and **Multiplicatio:** *The Yellowing or* **Citrinitas**

Figure 28 depicts the alchemical *opus* including the central role of Mercurius. Here, represented by a swan with outspread wings, as the 'Philosophick Mercury' supporting the union of the King and Queen, which

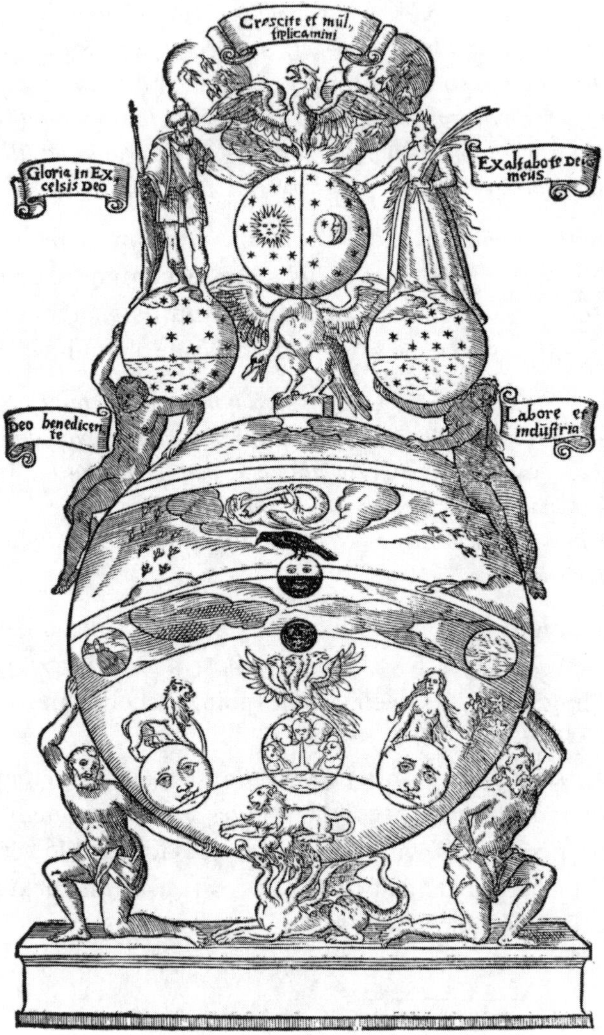

Figure 28. *Multiplicatio,* from *Alchymie,*
Andreas Libavius (1606), plate 15.

culminates in the *multiplicatio* stage, represented by the phoenix (a symbol for the *lapis)* on the upper sphere, cremating itself, and with multiple birds arising and flying out (in the *projectio* stage) from the ashes. Jung comments that this latter image of the multiple birds (coloured golden and silver) is the sign of multiplication and increase.[305]

Figure 29 (next page), though it has a strong Christian overlay, shows in another way how Mercurius is central to the work. He stands above and below, and both permeates and directs the *opus magnum.* The garden of the philosophers or sages is open (as the alchemical retort at the multiplication stage now can be, when the stone has become an inner, concrete and radiating reality). The garden is lined on three sides by a double avenue of grape vines. Under the earth the seven metals or Earthly Planets 'await their perfection, which is imparted to them by the perfection of the Work'.[306] Christ is the philosopher's stone in this image.

In alchemy, *multiplicatio* as the yellow stage is compared with the seeding or multiplication—with the grain of corn that multiplies in thousands and thousands. It represents a transforming emanation that extends to the environment, and even to the cosmos. Bees flying to and from their hive represent, in natural living form, this same unity in multiplicity. In an ideal form, the *multiplicatio* and *projectio* stage of the alchemical work would represent that in which a sage, a man or woman transformed through the deep alchemical process, could be constantly 'in tao'. This would mean everything within would correspond to and be identical with what is outside—in other words, there would be a synchronicity of events. In reality, however, this is a state into which we humans might fall only very briefly—it is the *unus mundus.*[307]

Andreas Schweizer says of the *citrinitas* or yellowing stage in alchemy):

---

305 *Psychology and Alchemy,* CW 12, par. 400, fig. 142.

306 From Steffan Michelspacher's *Cabala* (1616).

307 Von Franz, *Apuleius'* Golden Ass, X, p. 10.

Figure 29. *Multiplicatio* image,
from Steffan Michelspacher's *Cabala* (1616).

Psychologically speaking, we can say, that, for the first time, the yellowing embraces the awareness that the contents of the collective unconscious have a concrete, that is, substantial, reality. It is as if the first sunbeams in the morning throw their light onto the self. This light of consciousness leads to the cognition of an inner core of the psyche, by which the psychic life is regulated. Psychologically we call this inner core the self; the participants of a mystery celebration experience it as a revelation of the god-image, and the alchemists speak of the production of the gold. They all refer to one and the same unknown transcendental wholeness, which despite its transcendency, paradoxically manifests itself concretely, that is, in the realm of the sensible corporeal world.[308]

The solidity of the contained vessel is now able to become the solidity of the philosopher's stone—that is, a permanent inner experience of the Self that gives the personality innermost solidity, making an outer artificial solidity or solidification superfluous.[309] The vessel can now be opened and the new life flow or radiate out.

Another image to consider in the golden multiplication stage of the *opus* is related to the light of consciousness—sparks of light that radiate and illuminate. These *scintillae* (also often golden and silver) are found in multiple form in the earth, and are called 'fish eyes' by the alchemists. Of the fish's eye, Dorn says: this shining eye is the sun, which plunges 'the centre of its eye' into the heart of man 'as if it were the secret of warmth and illumination'. The fish's eye is always open like the eye of god.[310] The life force of eros is central in this process of illumination and multiplying healing energy that emerges and flows from the centre. The following sacred Prayer to Eros reveals something of the magnitude of the life force of the winged Eros.

I call to thee, the source of everything living, whose wings are outspread over the whole world, the unnameable and immeasurable, who breathes life-giving thoughts into every soul, who with his power has welded everything together. First born, all-creator, golden-winged, dark mysterious one, who conceals discreet and crafty thoughts and inspires dark and

---

308 A. Schweizer, lecture.

309 Von Franz, *Apuleius' Golden Ass,* X, pp. 8f.

310 *Mysterium Coniunctionis,* CW 14, par. 45.

ominous passion, thou hidden one who lives in stealth in every soul; thou kindlest the invisible fire in all that is animate by thy touch, tirelessly tormenting, with lust, through anguished rapture, since time began ... Thou youngest one, lawless, heartless, inexorable, invisible, bodiless, bringer of passion, archer, torch-bearer, thou master of every movement of the spirit, of all things hidden, giver of forgetfulness, father of silence, through whom and to whom the light shines forth, tender infant when you are born in the heart, ancient of days when you are quite completed. I call to thee the inexorable, using thy great name. Thou hast been the first to appear, nocturnal visitor, joyful by night, maker of night, thou who hears and answers ... thou in the depths ... thou hidden in the sea, thou eldest![311]

A Dream of Bees, Honey and the Hive

I will now include a relevant dream of mine:[312]

I am in the countryside at a university. It was like at UWS in Sydney [on the rural university campus which was a former Agricultural College and later home to leading bee-research work][313] *but I had a sense I was in*

---

311 'A Prayer to Eros', in von Franz, *Projection and Re-Collection in Jungian Psychology,* pp. 133f.

312 I dreamt this four months after I had returned to Australia, having lived for several months in Küsnacht, Zurich, where I had begun training at the C. G. Jung Institute.

313 I recently discovered a strong living connection with bees at this campus, which further amplified the dream's images for me. The international inaugural Eva Crane Memorial Award (2008) was won by a team led by Marc Greco of the Centre for Plant and Food Science based at the university in the dream—the University of Western Sydney in New South Wales, Australia. Dr Crane, OBE and founder of the International Bee Research Association, died in England, her home country, in September 2007. The award was instituted to celebrate her life and to mark her huge contribution to apicultural science for over fifty years. The award is made to 'the best, most innovative and scientifically exciting article published in the *Journal of Apicultural Research* each year, as judged by the current international panel of Editors.' (from 'Eva Crane Memorial Award 2008'; see www.ibra.org.uk. The winning article was 'X-ray computerised microtomography (MicroCT): a new technique for assessing external and internal morphology of bees'. MicroCT is emerging as a new method for the non-invasive imaging of insects. Greco has been developing Diagnostic Radioentomology (DR) to look inside insects without damaging them. Another remarkable factor in this work is that 'the non-destructive nature of DR will also be invaluable for rare or precious specimens or for non-invasive phylogenetic studies on ancient insects such as bees trapped in amber'. Greco's work also bridges Australia and Switzerland. At the time of publication he was

Zurich, attending to the enrolment details regarding my analyst training. I was standing at an outdoor kiosk waiting with others to see the course co-ordinator B. [a Professor of Psychology whose work includes an affinity with Analytical Psychology, Cultural Ecology and Buddhist studies] *to put my individual case to him.*[314]

There were students (mostly young 'earthy' males, perhaps agriculture students) standing beside me. Behind me and to the right was a low green wooden cupboard. It had doors that opened up and a lid that came right off so the top could be open to the sky. It was a beehive and big, healthy golden bees were on the outside working the honeycomb and there were some white cauliflower-like masses in the golden honey.[315] *I commented to a young man in his thirties: 'The bees are very healthy in this part of the world'.*[316] *At first I thought these big bees were the drones [the male bees that do no work, but mate with the queen]. I picked some of this white substance out of the honey and removed it from the honeycomb. I looked into the hive once or twice gently and was not stung. The bees were not aggressive. I could put my bare hand into the honey, which was about ten centimetres deep on the bottom of the cupboard. There were no moveable frames.*[317]

---

researching *Apis Mellifera* colony development at the Swiss Bee Research Institute. He is also a beekeeper and an expert on stingless Australian native bees.

314 I had spoken to B. in a similar context earlier on the day of the dream. As soon as I had completed my higher degree, I would be able to return to Zurich.

315 These were like bee brood from which new bees would develop. *Chou-fleur*, French for 'cauliflower', is associated with a term of endearment and part of the language of eros in French (*mon chou*: my love or darling). The white substances in a beehive include bee brood and larvae, as well as royal jelly.

316 The comment in my dream that 'the bees are very healthy in this part of the world' has turned out to be literally the case, as Australia is the only country in the world (as we approach 2011) that does not have the Varroa destructor mite and Colony Collapse Disorder in its bees. Australian bees, for example, have been sent en masse since 2006 to California to regenerate the seriously depleted bee populations and to help continue the pollination so critical to fruit and nut trees, and especially to the almond industry. The Australian-made film *Honeybee Blues* (released late 2009) is rich in information about the plight of the world's bee health and how the vitality of Australian bees is key to this.

317 In other words it was a wild or natural hive but was contained in this furniture that was like a European folk heirloom painted with flowers in folk-art style. It looked similar to the painted chest that my training analyst has in his practice room in Zurich, the room where I later found out that Jung had begun his work in private practice.

The setting of this dream was the university where I attended residential courses for my Masters degree, a prerequisite for the analyst training in Zurich. This campus was formerly an agricultural college. My stepfather Claude who, as I mentioned in the introduction, kept bees after he retired, used to run a hotel in a small country town where there was another agricultural college. Such colleges are not common. Claude was a widower then. He became like a therapist or father confessor to the young men who lived, studied and worked on the land there. He loved animals and had a dog and two 'Christian Cross' donkeys. These animals were real tricksters. The students would tell Claude their stories and their troubles; he would listen and was like a mentor for them. When I had this dream I was completing my higher degree so that I could return to Zurich where I had already begun my training analysis.

I was writing about Asclepius and the wounded healer archetype at that time. This had been a research topic of mine for several years. I was fortunate enough to be able to read and write about Jungian psychology for my Masters degree. For me, to be able to follow my individual love of analytical psychology was like being able to dip into the honey. It also linked me back to my own 'down-to-earth' work as a beekeeper, talking to farmers and working far from town and city life, and to my life with my stepfather who had died twenty years before. He was diagnosed with a virulent form of lung cancer and, while he was dying and after his death, I looked after the bees for him. I loved this work. I would have to wear overalls, a helmet and gloves and drive a truck to the bees. We would load them up and move them in the night to the next honey source, often in the eucalyptus forests. Once while on a mountain road driving a Landrover in the middle of the night, I remember being stung by stray bees trapped inside my veil. I did not ever curse the bees when they stung; I just removed the sting as cleanly and quickly as possible. I have always treated the bees with respect and I do not fear them in spite of experiencing many stings.

****

Another image and expression of the *multiplicatio* stage was the very old wise man in the Chinese film I saw in September 2006 at St Niklausen.

He had bees living in a cupboard in the kitchen of his hut. He was most delighted about this and his joy combined the openhearted delight of a child with the wisdom of old age. It seemed quite natural to him that these bees would live in his domestic space and be part of his daily life. He was very much at peace about the bees making their hive in his home and flying around as he cooked and went about his daily activities. Having a cupboard full of honeycomb, honey and bees seemed nothing out of the ordinary to him. His old face was radiant with joy and good nature. He smiled and chuckled as he showed the bees to his visitors. He could live side by side with this working beehive and had no thoughts of exterminating these honey-making little creatures that had the potential to sting him. He and they were 'in tao' together. The Berserker side of the bees was transformed and harmonious with this man and within the domestic container his humble home offered.

Towards the end of writing this book I was given a gift of a beautifully illustrated children's book called *The Bee-Man of Orn* (originally written in 1887). The story starts with a description of another such old man living in harmony with bees:

> *In the Ancient Country of Orn there lived an old man who was called the Bee-man, because his whole time was spent in the company of bees. He lived in a small hut which was nothing more than an immense beehive, for these little creatures had built their honeycombs in every corner of the one room it contained. . . . All day the air of the room was thick with buzzing insects, but this did not interfere in any way with the old Bee-man, who walked in among them, ate his meals, and went to sleep without the slightest fear of being stung. He had lived with the bees so long, they had become so accustomed to him and his skin was so tough and hard, that they no more thought of stinging him than they would of stinging a tree or a stone.*[318]

---

318 Frank R. Stockton, *The Bee-Man of Orn,* p. 1.

6
Conclusion

Deep eros is at the heart of my topic, and eros, like the bees, is a living mystery that cannot be defined. Marie-Louise von Franz, in *Projection and Re-Collection in Jungian Psychology,* asks the question: But who, even today, can really grasp the meaning of a deep experience of love?[319] Jung's well-remembered comments also bear witness to and express something of this living mystery:

> *Eros is a* kosmogonos, *a creator and father-mother of all higher consciousness. I sometimes feel that Paul's words—"Though I speak with the tongues of men and of angels, and have not love"—might well be the first condition of all cognition and the quintessence of divinity itself. Whatever the learned interpretation may be of the sentence "God is Love", the words affirm the* complexio oppositorum *of the Godhead. In my medical experience as well as in my own life I have again and again been faced with the mystery of love, and have never been able to explain what it is . . .*
>
> *For we are in the deepest sense the victims and the instruments of cosmogonic "love". I put the word in punctuation marks to indicate that I do not use it in its connotations of desiring, preferring, favouring, wishing, and similar feelings, but as something superior to the individual, a unified and undivided whole. Being a part, man cannot grasp the whole. He is always at its mercy. He may assent to it, or rebel against it; but he is always caught up by it and enclosed within it. He is dependent upon it and is sustained by it. Love is his light and his darkness, whose end he cannot see. "Love ceases not"—whether he speaks with the "tongues of angels," or with scientific exactitude traces the life of the cell down to its uttermost source. Man can try to name love, showering upon it all the names at his command, and still he will involve himself in endless self-deceptions. If he possesses a grain of wisdom, he will lay down his arms and name the unknown by the more unknown,* ignotum per ignotius*—that is, by the name of God. That is a confession of his subjection, his imperfection and his dependence; but at the same time a testimony to his freedom to choose between truth and error.*[320]

---

319 *Projection,* p. 133.

320 *Memories, Dreams, Reflections*, pp. 386f.

This wisdom of differentiated eros underlies the Transformed Berserker vision of Niklaus von Flüe. Elsewhere Jung writes:

> *What does lie within our reach . . . is the change in individuals who have, or create for themselves, an opportunity to influence others of like mind . . . anyone who has insight into his own actions, and has thus found access to the unconscious, involuntarily exercises an influence on his environment.*[321]

Conscious carriers of the opposites, as Edinger calls these individuals with insight into their own actions, are aware of the operation of the opposites within themselves, and have, to a greater or lesser extent, experienced the *coniunctio* (the union of psychic opposites).

> *To the extent that such individuals exist and carry the opposites within themselves, they do not feed the exteriorisation of the terrible strife between the opposites. There is reason to believe that if society is to be redeemed, it will be done through the cumulative effect of such people. And when a sufficient number of people carry the consciousness of wholeness, the world itself has a chance to become whole.*[322]

Marie-Louise von Franz also wrote of this happening in the particular case of the Transformed Berserker vision of Niklaus von Flüe, saying that even though 'we can only integrate our personal shadow, not the collective shadow of the Self, the dark side of the Godhead'—

> *Yet if we suffer the problem of the opposites to the utmost and accept it into ourselves, we can sometimes become a place in which the divine opposites can spontaneously come together. This is quite clearly what happened to Brother Klaus. His vision showed him that the divine opposites had become one within the Self and that this unified figure now overflowed with honey. This is a love that flows from the individual human being who has become whole.*[323]

Tolstoy, in the epilogue of his massive novel *War and Peace,* evokes

---

321 'The Undiscovered Self', *Civilization in Transition*, CW 10, par. 583.

322 Edinger, *The Mysterium Lectures*, pp. 325f.

323 Von Franz, 'The Transformed Berserker', in *Archetypal Dimensions of the Psyche,* pp. 48f.

the beehive with eros and openheartedness,[324] saying that human knowledge is indeed limited when we are confronted with the innate wisdom and power of nature (or, in psychological terms, of the unconscious or the objective psyche). The opposites of war and peace have become unified or combined here, just as they were in the image of the golden Transformed Berserker in the vision of Niklaus von Flüe. Tolstoy writes:

> *Just as the sun and each atom of ether is a sphere complete in itself, and at the same time is only a part of a whole inconceivable to man through its vastness, so every individuality bears within it its owns ends, and yet bears them so to serve general ends unfathomable by man.*
>
> *A bee settling on a flower has stung a child. And the child dreads bees, and says the object of a bee is to sting people. A poet admires the bee, sipping honey from the cup of the flower, and says that the object of the bee is to sip the nectar from the flower. A beekeeper, noticing that the bees gather pollen and bring it to the hive, says that the object of the bee is to gather honey. Another beekeeper, who has studied the life of the swarm more closely, says the bees gather honey to feed the young ones, and to rear a queen, that the object of the bee is the perpetuation of its race. The botanist observes that the bee flying with the pollen fertilises the pistil and in this he sees the object of the bee. Another, watching the hybridisation of plants, sees that the bee contributes to that end also, and he may say that the bee's object is that. But the final aim of the bee is not exhausted by one or another, or a third aim, which the human intellect is capable of discovering, the higher the human intellect rises in the discovery of such aims, the more obvious it becomes that the final aim is beyond its reach.*
>
> *All that is within the reach of man is the observation of the analogy of the life of the bee with other manifestations of life.*[325]

As Gotthilf Isler said in a lecture, 'there needs to be an acceptance of nature, of the feminine and a consideration of evil in our religious attitude,'[326] and that every individual has great importance and even a

---

324 More so than in his main religious work, *The Kingdom of God Is Within You,* as quoted here earlier.

325 Tolstoy, *War and Peace*, Epilogue. part 1, chap. 4, trans. Constance Garnett (1904). I chose this particular translation for its poetry. See also the inspiring new translation by Anthony Briggs (pp. 1269f.).

326 Isler, 'The Snake as Spirit of the Earth'.

divine aspect, also that everything depends on the sense of responsibility and integrity of each person. In the symbolism of the bee, the honey and the hive, the opposites are strong ('Out of the strong came sweetness' (Judges 14:14) or as alchemist Michael Maier wrote: 'Where honey is, there gall'),[327] but alchemically when the bright celestial spiritual masculine side descends to meet the dark but regenerative eros of the earthly feminine that ascends, a new life, a new springtime and dawn, as well as a living stone, an eternal resurrection body, is created. Life is alive and flowing, as well as radiating, from a centre or a source—a deep embodied wellspring—which affects the *anima mundi.* This is indeed something to rejoice in:

> *In the honey, the 'sweetness of earths', we can easily recognise the balsam of life that permeates all living, budding, and growing things. It expresses, psychologically, the joy of life and the life urge which overcome and eliminate everything dark and inhibiting. Where spring-like joy and expectation reign, spirit can embrace nature and nature, spirit.*[328]

Reflections

In the autumn of 2006, I revisited Jung's grave and those of Marie-Louise von Franz and her colleague Barbara Hannah. There were delicate pale pink waxy begonias with bright yellow centres flowering on the women's graves (they share one grave). Also the buddleia, so attractive to butterflies, was in bloom and many bees were happily working in the purple flowers. I found this reassuring for my writing on the bees. For a time I was alone and so sat quietly on the earth and watched the bees. The inscription on one tombstone said in Latin *Percussit petrum et fluxerunt aquae* ('Strike the stone and the waters will flow').[329]

I then remembered von Franz's beautiful dream in which the dreamer

---

327 Michael Maier, in *Symbola Aurea Mensae Duodecim Nationum* (1617), p. 515.

328 *Mysterium Coniunctionis,* CW 14, par. 698.

329 For more on the philosopher's stone and the water of life, see Edinger, *The Aion Lectures*, p. 75. In *Aion*, CW 9ii, Jung speaks of the water from the rock as the equivalent to the alchemical *aqua permanens*. 'Be you as living stones built up.' (I Peter 2:5) He goes on to say: 'the new rock [analogous to the philosopher's stone or *lapis*], then, takes the place of Christ, just as the everlasting Gospel was meant to take the place of Christ's message'.

+

finds the ordinary-looking mercurial water of life she was looking for in the apothecary's shop. The dream ends as she stands on the beach looking out at the sea with the man she knew she had always been married to and then she wakes with the expressed joyful realisation: 'That is the birth of Aphrodite'.[330] It celebrates the birth of Venus, goddess and queen of love who in turn gave birth to Eros (Cupid), the god of love with his golden arrows. Ares/Mars, the god of war, was his father, and Aphrodite also had an active interest in war and warriors. So once again we have love and war—the honey and the sting—together, and the cycle of their interplay goes on.

****

The inscription on Jung's tombstone (from I Corinthians 15:47), cradled by a small red-leafed maple tree over a little pool of water in a rock, reads:

> *Vocatus atque non vocatus Deus aderit.*
> *Primus homo de terra terrenus;*
> *Secundus homo de caelo caelestis.*
>
> [Called or not called, God will be present.
> The First Man is of the earth, earthly;
> The Second Man is of heaven, celestial.]

The living spirit of the unconscious brings a new consciousness and the water of life flows. I had the following dream when I was living in Küsnacht four years earlier. This was at the time that I took what turned out to be the first steps on my pilgrimage to St Niklausen, the place of the Transformed Berserker vision:

> *I was in Switzerland ... I was at my first meeting or analytic session with Marie-Louise von Franz. We met outdoors on a wooden platform or balcony like those jutting out onto the lake of Zurich in Küsnacht or like at the railway shelter at Schloss Laufen near the Rhine Falls, but more open to the air. I had not intended this, but realised there were about eleven people sitting there in the space. (They were my husband's family, perhaps some of his relatives who lived nearby in the Rhine Valley.)*

---

330 Von Franz, *C. G. Jung: His Myth in Our Time,* p. 207.

Von Franz was wearing a pale pink-apricot pastel coloured dress in a fine, soft flowing fabric like silk. It had a floral pattern of darker purple (Royal purple) and gold Heartsease flowers on it. At one point she stood up and walked over to be closer to me; she stood tall and had a real sense of authority about her. I apologised for all these people being here. I did not know why they had come; it was probably because they were part of the community and lived nearby.

At the end I had a sense of a square carved into wood (like an Asian wood carving; a finely chiselled pattern). There was also a sense of a square in a circle. The material was unpolished, unpainted wood with just some carving in relief on it. It was suspended (there was no sense of anything physical holding it up but it was being supported) behind where we were. It looked very natural.

The dream ends with the self-sustaining unity of the square and the circle. This is a mandala symbol of earth and the heaven meeting, and of wholeness. The dance movement and circumambulation of the bees is another form of this deeply nourishing and healing archetype of inner order and wholeness.

The life of the bees is a unique manifestation of circumambulation of both a queen and a wholeness—the hive. Numinosity and mystery remain in spite of the intense gaze of modern science, including neurobiology. The bees, in relation to the wholeness of their hive, express in their daily lives what Jung says, when speaking of the deity or the Self, 'the fact that there is a centre and a periphery, and it tries to embrace the whole. It is the symbol of wholeness.'[331] The life of the bees manifests this in so many ways, known and unknown.

---

331 'The Houston Films', in *C. G. Jung Speaking,* p. 327.

Acknowledgements

Heartfelt thanks to Rainer Rollfink and to Daryl Sharp. My deep appreciation goes to Andreas Schweizer.

And to Eva Wertenschlag-Birkhäuser, Hansueli Etter, and the *Stiftung für Jung'sche Psychologie,* Zurich, my gratitude for their appreciation of this work in its early stages. And to the Board of the Centre for Depth Psychology, Zurich.

The author would like to thank Lama Choedak Yuthok; Anne Baring and Jules Cashford; Andreas Schweizer; and Gene Robinson for permission to quote from their work. Thanks also to Princeton University Press for permission to quote from Eric Neumann, *The Origins of Consciousness;* and to Eva Crane and Hilda Ransome for their groundbreaking research in apiculture and bee lore.

(Every attempt has been made to trace and acknowledge copyright owners of material used in this book.)

Figure 30. Bees circumambulating the hive.

Figure 31. Danish woodcut showing hives under ridged roofs and beekeeping equipment (Hervigus, 1649). *Top left:* upright hives as used in Sweden; *to right:* horizontal hives as used in Denmark; *centre left:* two kinds of smoker; *centre right:* bees at water supply; *bottom:* four hives tools, a hive carried between two poles, with a square board, just above that, possibly for a swarm to settle on.

Bibliography

Abt, Theodor. *The Book of Pictures: Mushaf as-suwar by Zosimos of Panopolis*. Zurich: Living Human Heritage Publications, 2007.

Adler, Gerhard, *C. G. Jung Letters, vol. 1:1906–50*. Princeton, NJ: Princeton University Press, 1973.

Adler, Gerhard. *The Living Symbol: A Case Study in the Process of Individuation*. (Bollingen Series LXIII). New York: Pantheon Books, 1961.

Apollon: The Journal of Psychological Astrology. London: The Centre for Psychological Astrology, 1998.

Aristotle, 'Historia Animalium'. In *The Works of Aristotle*, vol. 4. Ed. J. A. Smith. Oxford: Clarendon Press, 1910.

Artis Auriferae quam Chemiam vocant antiquissimi authores, sive Turba Philosophorum [1572]. 3 vols. Basel: Apud Petrum Pernam, 1593.

Avalon, Arthur (Sir John Woodroffe). The Serpent Power: The Secrets of Tantric and Shaktic Yoga. London: Lazac & Co., 1919.

Atwood, M. A. *Hermetic Philosophy and Alchemy*. New York: Julian Press, 1960.

Bachofen, Johann Jakob. *Mutterrecht [Myth, Religion and Mother Right: An Investigation of the Religious and Juridical Character of Matriarchy in the Ancient World]*. Princeton, NJ: Princeton University Press, 1967.

Baring, Anne, and Cashford, Jules. *The Myth of the Goddess: Evolution of an Image*. London: Penguin, Arkana, 1993.

Berthelot, M. *Collections des anciens alchimistes grecs*, 3 vols. Reproduction of the 1888 (Paris) edition. Osnabrück: Otto Zeller, 1967.

Birkhäuser-Oeri, Sibylle. *The Mother: Archetypal Image in Fairy Tales*. Toronto: Inner City Books, 1988.

Bishop, Holley. *Robbing the Bees: A Biography of Honey the Sweet Liquid Gold that Seduced the World.* New York: Free Press, Simon & Schuster, 2005.

Bond, D. Stephenson. *The Archetype of Renewal: Psychological Reflections on the Aging, Death and Rebirth of the King*. Toronto: Inner City Books, 2003.

Brhadaranyaka Upanishad. Oxford, UK: Clarendon Press, 1900.

Briffault, Robert. *The Mothers: A Study of the Origins of Sentiments and Institutions*. London: George Allen & Unwin, 1927.

Burr, Elizabeth (trans.). *The Chiron Dictionary of Greek and Roman Mythology*. Wilmette, IL: Chiron Publications, 1994.

Buxton, Simon. *The Shamanic Way of the Bee: Ancient Wisdom and Healing Practice of the Bee Masters*. Rochester, VT: Destiny Books, 2004.

Cashford, Jules. *The Moon: Myth and Image*. New York: Four Walls, Eight Windows, 2003.

Cirlot, J. E. *A Dictionary of Symbols*, 2nd ed. London: Routledge, 1971.

Chetwynd, Tom. *Dictionary of Sacred Myth*. London: HarperCollins, 1986.

Choedak Yuthok, Lama. *Healing Relationships: Buddhist Insights for Enduring Happiness*. Melbourne: Lothian, 2005.

Columella, Lucius Julius Moderatus. *On Agriculture*. Revised ed. Trans. E. S. Foster and Edward H. Heffner. London: William Heinemann, 1968.

Cooke, Arthur Bernard. *Zeus: A Study in Ancient Relig*ion. London: Cambridge University Press, 1925.

Coomarawamy, Rama P. *The Door in the Sky*. Princeton, NJ: Princeton University Press, 1997.

Cooper, J. C. *An Illustrated Encyclopaedia of Traditional Symbols*. London: Thames & Hudson, 1978.

Crane, Eva. *A Book of Honey*. Oxford: Oxford University Press, 1980.

________. *The Archaeology of Beekeeping*. London: Duckworth, 1983.

________. *The World History of Beekeeping and Honey Hunting*. London: Duckworth, 1999.

Detienne, Marcel. *Dionysus at Large*. Cambridge, MA: Harvard University Press, 1989.

________. *Dionysus Slain.* Trans. Mireille Muellner and Leonard Muellner. Baltimore: Johns Hopkins University Press, c. 1979.

________. Orphée au miel. In *Faire de l'Histoire*, ed. Jacques Le Goff and Pierre Nora Paris: Gallimard, 1974.

________. *The Gardens of Adonis*: *Spices in Greek Mythology*. Princeton, NJ: Princeton University Press, 1977.

Dorn, Gerhard. *Congeries Paracelsicae chemicae de transmutation metallorum*. See *Theatrum Chemicum*.

du Bartas, Guillaume de Salluste. *Divine Weeks and Works*. [1578]. Trans. Joshua Sylvester. Gainesville: Scholars' Facsimiles and Reprints, 1965.

Dunne, Claire. *Carl Jung: Wounded Healer of the Soul, An Illustrated Biography.* Introduction by Jean Houston. New York: Parabola, 2000.

Ede, Piers Moore. *Honey and Dust: Travels in Search of Sweetness*. London: Bloomsbury, 2005.

Edinger, Edward F. *Anatomy of the Psyche: Alchemical Symbolism in Psychotherapy.* La Salle, IL: Open Court, 1985.

________. *Ego and Archetype*: *Individuation and the Religious Function of the Psyche*. Boston & London: Shambhala, 1992.

________.*Goethe's* Faust*: Notes for a Jungian Commentary*. Toronto: Inner City Books, 1990.

________.*The Aion Lectures: Exploring the Self in C. G. Jung's* Aion. Toronto: Inner City Books, 1996.

________. *The Mysterium Lectures: A Journey through C. G. Jung's* Mysterium Coniunctionis. Toronto: Inner City Books, 1995.

Edwardes, Tickner. *The Bee-Master of Warrilow*. London: Methuen, 1920.

Elder, John. 'Death of a Hero: Everest's Great Conqueror'. In the *Age*, 'Insight'. Melbourne, 12 January 2008.

Eliade, Mircea. *Images and Symbols: Studies in Religious Symbolism*. Princeton, NJ: Princeton University Press, 1991.

________. *Patterns in Comparative Religion*. New York: Sheed & Ward, 1958.

Evelyn-White, Hugh G. *Hesiod, The Homeric Hymns and Homerica*. [1914]. Cambridge, MA: Harvard University Press, 1964.

Evans, Arthur. 'The Palace of Knossos'. In the *Annual of the British School of Archaeology at Athens, VII*, 1900–1901.

Evans, Christine. Mysterious collapse of honeybee population threatens national food supply. www.naturalnews.com (21 March 2007).

Fife, A. E. The concept of the sacredness of bees, honey and wax in Christian popular tradition. Unpublished Ph.D. thesis. Stanford University, 1939.

Fludd, Robert. *Summum Bonum*. Frankfurt, 1629.

Gaster, Moses. *Rumanian Bird and Beast Stories*. London: Folklore Society, Sedgwick and Jackson, 1915.

George, Andrew, trans. *The Epic of Gilgamesh*. London: Penguin, 1999.

Gimbutas, Marija. *The Gods and Goddesses of Old Europe, 7000–3500 B.C. Myths, Legends, Cult Images*. London: Thames & Hudson, 1974.

________. *The Language of the Goddess*. San Francisco, CA: Harper and Row, 1989.

________. *The Living Goddesses*. Ed. and suppl. Miriam Robbins. Dexter. Berkeley, CA: University of California Press, 1999.

Goodman, L. J. and Fischer, R. C., eds. *The Behaviour and Physiology of Bees.* Wallingford: CAB International on behalf the International Bee Research Association, 1991.

Gould, James L. and Gould, Carol Grant. *The Honey Bee.* New York: Scientific American Library, 1988.

Grimal, Pierre. *The Dictionary of Classical Mythology*. Oxford: Blackwell, 1986.

Grimm Brothers. *Complete Grimm's Fairy Tales.* London: Routledge & Kegan Paul, 1975.

Grossman, David. *Lion's Honey: The Myth of Samson.* Melbourne: The Text Publishing Company, 2005.

Gruppe, Otto. *Griechische Mythologie und religionsgeschichte*. Munich: Beck, 1906.

Hall, James. *Hall's Illustrated Dictionary of Symbols in Eastern and Western Art*. London: John Murray Publishers, 1994.

Harrison, Jane. *Themis: A Study of the Social Origins of Greek Religion*. Cambridge: Cambridge University Press, 1912.

Henderson, Joseph, and Sherwood, Dyane. *Transformation of the Psyche: The Symbolic Alchemy of the* Splendor Solis. Hove, UK: Brunner Routledge, 2004.

Henry, Elizabeth. *Orpheus with his Lute: Poetry and the Renewal of Life*. Bristol: Bristol Classics Press, 1992.

Hesiod. See Evelyn-White, Hugh G., trans.

Hine, Daryl, trans. *Theocritus: Idylls and Epigrams*. New York: Atheneum, 1982.

Hobley, Brian. 'The Point in the Circle: The Image of Apollo in the Greco-Roman World'. In *Apollon: the Journal of Psychological Astrology*. Issue 1, October 1998. London: CPA Press.

Homeric Hymns, The. See Evelyn-White, Hugh G.

Hiller von Gaertringen, Friedrich. ‘Aristaios’. In G. Wissowa et al. eds., *Paulys Real-Encyclopädie der classischen Altertumswissenschaft.* Stuttgart, 1894–.

Hughes, Ted. *Birthday Letters.* London: Faber and Faber, 1998.

International Bee Research Association. Eva Crane Memorial Award 2008. Published 19 February 2009. www.ibra.org.uk.

Isler, Gotthilf. ‘The Snake as Spirit of the Earth’. Lecture, September 2004, St Niklausen, Obwalden, Switzerland, 2004.

Jaffé, Aniela. *An Archetypal Approach to Death Dreams and Ghosts.* Einsiedeln: Daimon Verlag, 1999.

Johnson, Buffie. *Lady of the Beasts: The Goddess and Her Sacred Animals.* San Francisco: Harper and Row, 1988.

Jonson, Ben. *The Alchemist.* Performed Oxford, 1610; printed 1613.

Jung, C. G. *The Collected Works* (Bollingen Series XX). 20 vols. Trans. R.F.C. Hull. Ed. H. Read, M. Fordham, G. Adler, Wm. McGuire. Princeton, NJ: Princeton University Press, 1953–1979.

________. *C. G. Jung Letters,* vol. 2, 1951–1961. (Bollingen Series XCV-2). Princeton, NJ: Princeton University Press, 1953–1975.

________. *Memories, Dreams, Reflections.* London: Routledge and Kegan Paul, 1963. Revised edition: New York, Random House (Vintage Books), 1989.

________. *Nietzsche's* Zarathustra*: Notes on the Seminar given in 1934–1939.* Princeton, NJ: Princeton University Press, 1988.

________ et al. *Man and His Symbols.* London: Arkana, Penguin Books, 1990.

Komarios. ‘Komarios to Cleopatra’. In Berthelot, M., *Collections des anciens alchimistes grecs*, vol. 2. Reproduction of the 1888 (Paris) edition. Osnabrück: Otto Zeller, 1967.

Kerényi, Karl. *Apollo: The Wind, the Spirit, and the God: Four Studies.* Dallas, TX: Spring Publications, 1983.

________. *Asklepios: Archetypal Image of the Physician's Existence*. New York: Bollingen Foundation, Pantheon Books, 1959.

________. *Dionysos: Archetypal Image of Indestructible Life*. London: Routledge and Kegan Paul, 1976.

________. *The Gods of the Greeks*. London: Thames & Hudson Ltd, 1951.

Kidd, Sue Monk. *The Secret Life of Bees*. London: Viking Penguin, 2001.

Klossowski de Rola, Stanislav. *The Golden Game: Alchemical Engravings from the Seventeenth Century*. London: Thames & Hudson, 1988.

Langstroth, Lorenzo Lorraine. *A Practical Treatise on the Hive and the Honey Bee*. New York: C. M. Saxton & Co., 1853.

Lockie, A., and Geddes, N. *The Complete Guide to Homeopathy*. London: Dorling Kindersley, 1995.

Libavius, Andreas. *Alchymia*. Frankfurt: printed by Joannes Saurius for Petrus Kopffius, 1606.

Luke, Helen. *Dark Wood to White Rose*. Pecos, NM: Dove Publications, 1975.

Macey, Richard. 'The buzz in science is what makes a bee so smart'. In *Sydney Morning Herald*, 17 October 2006.

Maier, Michael. *Symbola aurae mensae*. Frankfurt: at the press of Antonius Hummius for Lucas Jennis, 1617.

________. *Tripus aureus*. Frankfurt: at the press of Paul Jacobi for Lucas Jennis, 1618.

Mandeville, Bernard. 'The Grumbling Hive or Knaves Turn'd Honest'. In *The Fable of the Bees, or Private Vices, Publik Benefits* [1714]. F. B. Kaye (ed.). Indianapolis: Liberty Fund, 1988.

Melville, Kate. Social interactions alter gene expression. Urbana-Champaign, IL: University of Illinois, 2008.

Michelspacher, Steffan. *Cabala*. Augsburg: at the press of David Francken, 1616.

Meier, C.A. *Healing Dream and Ritual: Ancient Incubation and Modern Psychotherapy*. Einsiedeln: Daimon Verlag, 1989.

________. The Psychological Parameters of the Examination Situation'. In *The Well-tended Tree: Essays into the Spirit of Our Time*. New York: G. P. Putnam's Sons, for the C. G. Jung Foundation for Analytical Psychology, 1971.

Maeterlinck, Maurice. [1901]. Trans. Alfred Sutro. *The Life of the Bee*. London: The Folio Society, 1995.

March, Jenny. *Cassell's Dictionary of Classical Mythology*. London: Cassell, 1998.

Michelet, Jules. *L'Insecte*. Paris: Calman-Levy, 1858, reprinted 1903.

Mills, John. *An Essay on the Management of Bees*. London: J. Johnson and B. Davenport, 1766.

Musaeum Hermeticum. Frankfurt, 1678. See also Arthur Edward Waite, trans., *The Hermetic Museum restored and enlarged*, 3 vols., London: 1893.

Nature: International Weekly Journal of Science 443, 26 October 2006.

Neumann, Eric. *The Great Mother*. New York: Pantheon, 1955.

____________. *The Origins and History of Consciousness*. Princeton, NJ: Princeton University Press, 1954.

Newman, Leonard Hugh. *Man and Insects: Insect Allies and Enemies*. Garden City, NY: National History Press, 1965.

Otto, Walter. *The Homeric Gods: the Spiritual Significance of Greek Religion*. London: Thames & Hudson, 1954.

Ovid. *Fasti*. Trans. A. J. Boyle and Woodward, R. D. London: Penguin, 2000. Also trans. Anthony S. Kline, www.poetryintranslation.com

Paracelsus. *Selected Writings*. 2nd ed. Ed. Jolande Jacobi. New York: Pantheon Books, Bollingen Foundation Inc., 1958.

Pausanius. *The Description of Greece*. Trans. W. H. S. Jones and H. A. Ormerod. Cambridge, MA: Harvard University Press, 1918.

Penotus, Bernard. *Characteres secretorum celandorum*. In *Theatrum Chemicum*. London: printed for John Wyatt at the Rose in St Paul's Churchyard, 1692.

Philostratus the Elder, *Vita Apollonius Tyanaeus. Philostratus: The Life of Apollonius of Tyana*. Ed. Christopher P. Jones. Cambridge, MA: Harvard University Press, 2005.

Plath, Sylvia. *Ariel*. London: Faber and Faber, 1965.

Pliny the Elder, *Natural History*. 10 vols. Trans. H. Rackman. Revised ed. London: William Heinemann, 1968.

Porphyry. *The Cavern of the Nymphs*. Trans. Robert Lamberton. Barrytown, NY: Station Hill Press, 1983.

Purchas, Samuel. *A Theatre of Politicall Flying-Insects*. London: R. J. Parkhurst, 1657.

Preisendanz, K. *Papyri Graecae Magicae*. 2 vols. Stuttgart, 1973.

Ransome, Hilda M. *The Sacred Bee in Ancient Times and Folklore*. Mineola, NY: Dover Publications, 2004. Originally published, London: George Allen & Unwin, 1937.

Rig-Veda. Trans. Ralph T. H. Griffith. Delhi: 1973.

Roberts, Allen. *The Craft and Its Symbols*. Richmond, VA: Macoy Publishing, 1974.

Robinson, Gene et al. The Honey Bee Genome Sequencing Consortium. Proposal for the Sequencing of a New Target Genome: White Paper for a Honey Bee Genome Project. Represented by the authors of this document: G. E. Robinson, B. K. Aronstein, J. E. Evans, S. E. Fahrbach, J. S. Johnston, R. Maleszka, R. E. Page, H. M. Robertson, D. B. Weaver. generobi@life.uiuc.edu, 2002.

Rosarium philosophorum. See *Artis auriferae*.

Scheinberg, Susan. 'The Bee Maidens of the Homeric Hymn to Hermes'. In *Harvard Studies in Classical Philology 83*, 1979.

Schweizer, Andreas. (Lecture). The Teaching of Komarios and Cleopatra: Reflections on a Greek alchemical text dedicated to Komarios and Cleopatra, Zurich: ISAP, 2007. Also 'Observe Nature and You Will Find the Stone: Reflections on the Alchemical Treatise "Komarios to Cleopatra" '. In *Spring 82*, Fall 2009. New Orleans, LA: Spring Journal and Books.

Schweizer, Regine. (Lecture). The Divine Girl. St Niklausen, Switzerland, 2005.

Seeley, Thomas D. *The Wisdom of the Hive: The Social Physiology of Honey Bee Colonies*. Cambridge, MA: Harvard University Press, 1995.

Seneca. *Moral and Political Essays*. Trans. and ed. John M. Cooper and J. F. Procopé. Cambridge: Cambridge University Press, 1995.

Serrano, Miguel. *C. G. Jung and Hermann Hesse: A Record of Two Friendships*. London: Taylor and Francis Books, 1966.

Singer, June. *Boundaries of the Soul: The Practice of Jung's Psychology*. New York: Anchor Books, 1973.

Stein, Murray. *Jungian Analysis*. 2nd ed. La Salle, IL: Open Court, 1995.

Steiner, Rudolf. *Bees: Lectures by Rudolf Steiner*. Great Barrington, MA: Anthroposophic Press, 1998.

Stevens, Anthony. *Ariadne's Clue: A Guide to the Symbols of Mankind*. Princeton, NJ: Princeton University Press, 1998.

Stockton, Frank R. *The Bee-Man of Orn*. [1887]. Revised edition, illustrated by P. J. Lynch. Cambridge, MA: Candlewick Press, 2003.

Swammerdam, Jan. *The Book of Nature or the History of Insects*. Trans. Thomas Flloyd. London: C. G. Seyffert, 1758.

Theatrum Chemicum, 3 vols. I–III, Usel, 1602. IV–VI, Strasbourg, 1613, 1622, 1661.

Theocritus. Love Stealing Honey, *Epigram* XIX. See, *Theocritus: Idylls and Epigrams*, Daryl Hine, trans.

Trismosin, Salomon. *Splendor Solis*, MS Harley 3469 [1582]. Trans. J. K[ohn]. London: Kegan Paul, Trench, Trubner & Co, n.d. (ca.1920).

Tolstoy, Leo. *The Kingdom of God is Within You*: *Christianity Not as Mystic Religion but as a New Way of Life.* Trans. Constance Garnett. London: Heinemann, 1894.

__________. *War and Peace*. Trans. Constance Garnett. London: Heinemann, 1971.

Valentino, Basil. *Tripus aureus*. Illustrated by Michael Maier. Frankfurt, 1618.

Van der Post, Laurens. *Jung and the Story of Our Time*. New York: Random House, 1975.

________. *The Heart of the Hunter*. London: Vintage, 2002.

Van Over, Raymond, ed. *Eastern Mysticism*, vol. 1. 'The Near East and India'. New York: Mentor/New American Library, 1977.

Varro, Marcus Terentius. *De Re Rustica (On Farming)*. Trans. and ed. Lloyd Stoor-Best. London: G. Bell and Sons, 1912.

Virgil, *The Aeneid.* Trans. and ed. James Lonsdale and Samuel Lee. In *The Works of Virgil.* London: Globe Library, Macmillan, 1871.

________. *The Georgics*. Trans. Cecil Day-Lewis. London: Jonathon Cape, 1940.

Von Franz, Marie-Louise. *A Psychological Interpretation of* The Golden Ass *of Apuleius*. New York: Spring Publications, 1970.

________. *Alchemy: An Introduction to the Symbolism and the Psychology*. Toronto: Inner City Books, 1980.

________. *Archetypal Dimensions of the Psyche*. Boston: Shambhala, 1999.

________. *Archetypal Patterns in Fairy Tales*. Toronto: Inner City Books, 1997.

________. *Aurora Consurgens*. Toronto: Inner City Books, 2000.

________. *C. G. Jung: His Myth in Our Time*. New York: G. P. Putman's Sons, 1975.

________. *Creation Myths*. Dallas, TX: Spring Publications, 1972.

________. Interview with Suzanne Wagner, 1977. Transcript published in *Psychological Perspectives*, no. 38 (Winter 1998–9).

________. *On Dreams and Death: A Jungian Interpretation*. Peru, IL: Open Court, 1998.

________. *Projection and Re-collection in Jungian Psychology: Reflections of the So*ul. Peru, IL: Open Court, 1980.

________. *Shadow and Evil in Fairy Tales*. Boston: Shambhala, 1974.

________. *The Cat: A Tale of Feminine Redemption*. Toronto: Inner City Books, 1999.

________. *The Feminine in Fairy Tales*. New York: Spring Publications, 1972.

________. *The Interpretation of Fairy Tales*. Boston: Shambhala, 1996.

________. *The Passion of Perpetua: A Psychological Interpretation of Her Visions*. Toronto: Inner City Books, 2004.

von Frisch, Karl. *Bees, Their Vision, Chemical Senses and Language*. Ithaca, NY: Cornell University Press, 1950.

________. 'Decoding the Language of the Bee'. Nobel Lecture, 12 December 1973. In *Nobel Lectures Physiology or Medicine 1971–1980*. Ed. Jan Lindsten (Karolinska Institute, Stockholm). Singapore: World Scientific Publishing Co., 1992.

________. *The Dance Language and Orientation of Bees*. Cambridge, MA: Harvard University Press, 1967.

Waite, Arthur Edward, trans. *The Hermetic Museum, restored and enlarged.* 3 vols. London, 1893.

Wallis Budge, Ernest Alfred. *The Book of the Opening of the Mouth*, vol. ii. London, 1901.

Wilhelm, Richard. *The* I Ching *or Book of Changes*. Trans. Cary F. Baynes. Princeton, NJ: Princeton University Press, 1967.

Wilhelm, Richard, trans. The *Secret of the Golden Flower*. New York: Routledge and Kegan Paul, 1931.

Whitmont, Edward. *Psyche and Substance—Essays on Homeopathy in the Light of Jungian Psychology*. Berkeley, CA: North Atlantic Books, 1980.

________. *The Symbolic Quest: Basic Concepts of Analytical Psychology*. Princeton, NJ: Princeton University Press, 1991.

________. *The Alchemy of Healing: Psyche and Soma*. Berkeley, CA: North Atlantic Books, 1993.

Wilson, Bee. *The Hive: The Story of the Honeybee and Us*. New York: Thomas Dunne Books, St Martin's Press, 2004.

Worrel, Thomas D. (Lecture) 'The symbolism of the Beehive and the Bee'. Mill Valley Masonic Lodge, 29 February 2000.

Yeats, William Butler. 'The Second Coming' [1919]. In *Collected Poems of W. B. Yeats*. 2nd ed. London: Macmillan & Co., 1950.

Zagorski, Nick. 'Profile of Gene E. Robinson'. *Proceedings of the National Academy of Science* (PNAS), vol. 103, no. 44. Washington, DC: National Academy of Sciences of the USA, 2006.

Zimmer, Heinrich. *Philosophies of India*. New York: Bollingen Foundation, and Princeton, NJ: Princeton University Press, 1951.

Index

Entries in *italics* refer to illustrations

Abt, Theodor: *The Book of Pictures,* 95n, 128, 137
Acteon, 59
Adonis, gardens of, 90, 118
albedo, 32, 116, 131
alchemy, 9-10, 14, 24, 27, 31-32, 41, 50, 55, 63, 66-69, 75, 80, 95n, 97, 100, *103,* 110, 113-118, 124, 129,130-169
  and individuation process, 10, 24, 116, 130-169
  and well-sealed vessel, 23, 25, 63, 132, 147, 165
allopathic medicine, 78
altruism, 19, 21, 37, 40, 134n, 135
Ambrose, Saint, 64
Amerindians, 89
anima, 62, 131-132, 144n
anima mundi, 62, 159, 173
animus, 131-132
Aphrodite, 65, 113, 119, 174
anthropos, 108, 128, 142
apiarists, *see* beekeepers
apiary, 13, 26, 34, 36, 37, 38, 39, 138
Apis mellifera, 14, 18, 40, 135
Apollo, 58-60, 64-66, 69, 71, 72-78
aqua permanens, 77, 78, 144, 150, 173n. *See also* water(s)
archetypal/archetype(s), 10, 41-42, 49, 75, 114, 132, 149, 151n, 157, 168,
  of inner order, 9-10, 31-37, 40, 49, 65, 94, 101, 114, 118, 127, 151, 157, 175. *See also* mandala; Self
Aristaeus, 38, 53, 57-60, *61,* 62, 63, 79
Aristotle, 51, 79
arrow(s), 9-10, 73, 84 119-121, 174
Artemis, 51, 65-66, 125n
ascent and descent, 20, 56, 66, 72, 79, 150, 152, 153, 158
Asclepia/Asclepius, 72, 78, 89, 168
asexual worker bees, 18-19, 113
Assumption of Virgin Mary, 100
Astarte, 113
Astraea, 115
astrology, 65, 73n, 78, 84, 114-115, 148, 152, 156
atman, 108, 132, 133
Athena, 48
Attis, 90n, 118n
Auxo, 60
axis mundi, 66

Bachofen, J. B., 50-51
balsam (honey), 8, 143, 155, 159, 173
Baring, Anne, 49, 107n
Bassus, Cassianus, 63
bear. *See* berserk/Berserkers
  drowning bees, *126*
bee(s), *passim*
  'age castes', 21
  and alchemy, 9, 27, 31, *103,* 130-169
  anatomy of, *17, 120*
  and Aristaeus, 57-63
  and Cupid's arrows/stinging, 9-10, 119-121
  asexual females, 18-19, 113
  bee Sura, 104
  in Bible, 95-96
  black and golden sides, 81-82
  in 'bow of love', 105

bees(s) *(cont,):*
blue, 105
brain, 40, 44
and Buddhism, 109-111
bugonia, 53, *54,* 57-63, 96, 147, 148, 154
and bull, 47, 53, 62-63, 65, 66, 148, 154
'bull-born', *see bugonia*
calculating distance, 28, *30,* 43
castes, 11, *17,* 21
and cathedral builders, 102-104
centring, 151-152
and Christ/Christianity, 64, 96-102, 126, 128, 132, 151, 163
circumambulating the centre, *passim,* 10, 12, 45, 65-66, 68, 69, 87, 130, 135, 151-154, 156, 175, *176*
and Colony Collapse Disorder, 11, 13n, 36-39, 44, 62n, 167n
and colour vision, 41-42
and consciousness, 23-24, 27, 94
and creativity, 23, 83
cult, 47-49
and Cupid's arrows, 9, 119-121
dance language, 24, 27-28, *29-30,* 43, 149, 151, 175
and devil, 79, 81-83
differentiation of tasks, 19-20
and divination, 64-67
dream(s) of, 166-169
drones, *17,* 18-19, 21, 32, 50, 53, 65, 73n, 83-84, 137, 138-139, 140, 141
drowned by bear, 125, *126*
and Eros/Cupid, 9-10, 65, 119, 170
as 'eusocial' insect, 23, 135
flapping wings, 26
as foragers, 19, 21-22, 28, 43, 135, 151
and Freemasons, 102-103
genome, 9, 14, 22, 24, 37, 38, 40, 44
Goddess, 47-54, 66, 89
and Great Mother, 27, 47-54, 118, 141
hieroglyphics, *70*
and Hinduism, 105-109
and hive, 18-23, 46, 104, 115-116, 130, 169, *177. See also* hive
and honey. *See* honey
industriousness, 9, 10, 17-19, 39, 79, 82-83, 102-104, 116
and inner order, 9-10, 31-37, 40, 49, 65, 94, 101, 114, 118, 127, 151, 157, 175
interregnum, 32-36, 127, 139. *See also* swarm/swarming
and Islam, 104-105
jewellery, *47-49*
and Judaism, 95-96
and Kama, 105, *106*
and Karl von Frisch, 28, *29-30,* 38, 41-45
and Kundalini, 26-27
and love, 100, 108, 112-113, 118-119. *See also* eros *and* deep eros
maidens, 73-79
Mercurial nature of, 81-83, 131
as messengers of the gods, 64, 79-83, 87, 91, 131
and motherhood, 51
in nature, 17-45, 105
and nectar, 9-11, 14, 16-17, 19, 22-28, 30, 42-43, 50, 56, 58, 87-88, 91, 105-106, 110, 116, 129, 151, 172
as nurses, 19-20, 51

numinosity of, 9, 15, 18, 39, 45-46, 66-69, 80, 86-87, 96, 135, 138, 151, 175
and mandala symbolism, 10, 101, 152-157
and Muses, 10, 64-65
Mystery of the Queen, 40, 45-46, 135, 153, 175
navigation and dance language, 24, 27-28, *29-30,* 44, 77, 149, 151
and nectar, 14, 17, 23-26, 30, 43, 50
nigredo, 31-32, 75, 116, 131, 142, 154
on *omphallos* stone, *67*
oracle and Muse, 10, 11, 64-67
and pollen, 13-16, 19, 22-25, 27-28, 30, 43n, 89, 110, 172
as pollinators, 11, 13, 14, 36-38, 46, 50, 80, *99, 102,* 130, 131, 146
priestesses, 51, 65
psychological perspective, 66, 112-129
as psychopomp, 78
queen, 12-13, *17,* 18-23, 83, 140-141. *See also* queen bee
and Re's tears, 91-94
and the religious, 86-111
and salt, 42-43
as scouts, 27-28, 32
as sentinels, 19-20
sexuality of, 113-11, 131
and smoke, 31
as 'stars of heaven', 89
sting/stinging, 16, 22, 31, 46, 55, 83, 87, 109, 115, 118-119, *120,* 121, 126, 131, 134n, 174
and arrows, 119
swarm/swarming, 32-36, *35,* 71, 127, 131
symbolism in myth, legend and story, 10-11, 46-85
and tapas, 24-25
task differentiation in, 19-20, 30, 135
and Trophonius, *71,* 72-73
types of, *17*
undomesticated, 16
'unity in multiplicity,' 21, 135, 159
'veil-winged', 65-66
venom, 9, 22, 40
and virginity, 51, 114
and wasps, 41-42
and wax, 12, 19, 22, 24-26, 31, 66, 72, 83, 90, 93-97, 137, 140
as weather prophets, 79
wintering, 31-32, 131
workers, *17,* 18-19, 21, 26-30, 33, 50, 84, 118

Bee Goddess, 47-54, 66
beehive, *see* hive
beekeepers, 13-16, 32-*35,* 110, *177*
death of, 39
three archetypal kinds of, 38-39
Bee-Man of Orn, The, 169
'Bee Prince', 69n
beeswax, 12, 19, 22, 24-26, 31, 66, 72, 83, 90, 93-94, 97, 137, 140
uncapping, 12, 31
Benedictio Fontis, 97
Berenbaum, May, 38
Ber-serk, 125
berserk/Berserkers, 125-126. *See also* Transformed Berserker
Birkhäuser-Oeri, Sybille: *The Mother: Archetypal Image in Fairy Tales,* 91

bitterness, 10, 41-43, 96, 118, 173
blood, 36, 50, 55-56, 63, 74, 119, 129, 137, 138, 142, 156
sacrifice, 74
blue, 15, 42, 105, 123, 124, 128, 138, 150
alchemical, 137, 156n, 158, 159, 160. *See also* caelum
Bond, D. Stephenson: *The Archetype of Renewal,* 34n
'bow of love', 105
Brother Klaus, *see* Von Flüe, Niklaus
brood, 12, 13, 23, 25-26, 32, 167n
Browning, Zac, 37
bugonia, 53, *54,* 57-63, 65, 147, 154
Buddha nature, 109
Buddhism, 80-81, 109-111
bull worship, 47
'bull-born' bees, *see* bugonia
Buxton, Simon: *The Shamanic Way of the Bee,* 18

caelum, 69, 137, 148, 155-162, 156n
Calliope, 58
Cashford, Jules, 49
castes of bees, *17*
cathedral builders, 102
Cecrops, 72
Celtic druids, 89
centring, 151-152. *See also* circumambulating the centre
chaos, 10, 32-36, 80, 118
chastity, 21, 97. *See also* virgiinity
chthonic, 56, 65, 72, 79, 86, 113
chthonii, 72
Chiron, 58-60
Christ/Christianity, 64, 96-102, 126, 128, 132, 151, 163
chrysopoeia, 9, 130, 132
circulatio, 75, 122, 147, 151-156
circumambulating the centre, *passim,* 10, 12, 45, 66, 69, 87, 109, 122, 130, 135, 151-154, *176*
citrinitas, 162-166
Colony Collapse Disorder, 11, 13n, 36-39, 44, 62n, 167n
Columella, 10-11
community of the Self (spirit), 12, 46, 68, 130, 134. *See also* ecclesia spiritualis
coniunctio, 148, 155-162
of 2 and 3, 13n, 171
conjunction, 134-137, 148, 171. *See also* coniunctio
consciousness, 23-24, 31-32, 43-44, 73, 94-95, 124, 147, 151n, 165. *See also* ego; logos
lunar, 27, 38n, 73n, 77-78, 118
Crane, Eva: *A Book of Honey,* 91-92
Memorial Award, 166n
creation, 150, 152, 154, 158, 159
creation myths and folk stories, 80
Cretans/Crete, 49-50, 55-56, 89. *See also* Minoans
creativity, 31, 83-84, 116-117, 161
crisis, psychological, 117, 124, 127
Cupid, 84, *85,* 106, 119-121, 174. *See also* Eros

dance language, 27-28, *29-30,* 43, 77, 149, 151, 175
Danish hives and beekeeping equipment, *177*
darkness. *See* nigredo; wintering
deep eros, 11, 57, 112, 121, 123-124, 127-128, 143, 144-146, 150, 170

Delphi and divination, 65-68
Delphic Bee, 65
Demeter, 51, 65
gardens of, 90
Demeter-Ceres-Spes, 27, *52*
Detienne, Marcel: *Dionysos Slain,* 74
diamond body, 80, 109-110, 132
Diana, 66
differentiated eros, 124, 171
differentiation, 90
Dike, 115
Dionysos, 51, 55-59, 74, 79, 89
direction finding, *27-30. See also* navigational skills
disorientation. *See* interregnum; swarm/swarming
divination, 60, 64-67, 73-79
divine inspiration, 71, 124
domestication, 38-39
Dorn, Gerhard, 129, 155-165
dream(s), 45, 69, 78, 93, 151n, 160
of bees, honey and the hive, 166-169
Jung's, when writing *Psychological Types,* 117
of travelling to St Niklausen, 12
of 'garment of white' (solification garment) 145
of water, clear and flowing, 150
drones, *17,* 18-19, 21, 32, 50, 53, 65, 73n, 83-84, 131, 137-141
drought, 79, 150, 151n
Du Bartas, Guillaume, 21

ecclesia spiritualis, 50, 127, 130
Edinger, Edward F.: *Anatomy of the Psyche,* 23, 153
The Mysterium Lectures, 96, 161, 171
Edwardes, Tickner, *The Beemaster of Warrilow,* 39
ego, 37, 45, 77, 108, 134, 137, 146-147, 151-153, 157
and Self, 45
Egypt/Egyptian, 15n, 62-63, 88, 91-95
bee hieroglyphics, *70*
Goddess Nut, 91, *92*
mummification, 93
'Opening of the Mouth' Ceremony, 91
pharoahs, 68n, *70,* 94
Re, 69, 80, 92-94
royal names, 68n
Sed festival, 34
sun god's birth from lotus, *81*
Einstein, Albert, 37
Eleusinian Mysteries, 89
embodied spirit, 46, 124
Erechtheus, 72
Eros/Cupid, 9-10, 84, *85,* 118-121, 165-166, 170, 174
eros, 11, 39, 49, 55, 57, 75, 96, 100, 106, 118, 127, 143, 144-146. *See also* deep eros
differentiated, 124
and *multiplicatio,* 147, 162-166
essenes, 65
Eucharist, 89
Eurydice, 60, 62
examen/examinations, 34
'Eye of Heaven', 69
'Eye of the Universe', 69

fairy tales. *See Grimm Brothers*
feminine, 27, 60, 64-65, 75, 78, 80, 113, 125n, 141, 147, 150, 153, 158, 172-173
potency and fertility, 47-54, 147
Fifth Brahmana, 107-108
filius philosophorum, 33, 143

flowers (flowering), 11, 14, 15, 25, 30, 41, 42, 50, 51, *71, 80,* 81, 95, 100, 104, 105, 109, 110, 112-114, 118, 130, 131, 142, 146-150, 160, 161, 167n, 173
forager bees, 19, 21-22, 28, 43-44, 135, 151
foundation honeycomb, 12n, 13, 97, 119
Freemasons, 102-104
fruit(ing), 14n, 15, 20, *71, 80,* 104, 110, 124, 130, 163

Gaia, 58, 60
Galician Madonna and child, *98*
genome, 9, 14, 22, 24, 37, 38, 40, 44
Gimbutas, Marija, 47, 53
glutinum mundi, 23
Goddess of regeneration, *47,* 53
god-image, 10, 101, 113, 115, 127-128, 165. *See also* Self
gold, 9, 11, 14n, 24, 48, 75, 126, 130, 131, 141-144, 147, 154, 165
'Golden Flower', 25, 81, 133, 152, 161
Great Goddess, 27, 147
Great Mother, 46-54, 65, 78, 89, 131, 147
Greco, Marc, 166-167n
Greene, Liz, 73n
  The Astrology of Fate, 115
Grimm Brothers: 'Lazy Harry', 19n, 90
  'The Queen Bee', 20, 23, 57
  'The Valiant Little Tailor', 23
grounding, 117-118
Gruppe, Otto: *Griechische Mythologie,* 79

Hannah, Barbara, 173
healing, 9, 11, 14, 20-26, 33, 40, 44, 59-60, 67, 71-72, 75, 78-79, 88-89, 96, 105, 110, 112, 115, 118-122, 124, 127, 129-131, 135, 140, 143, 148-9, 165, 168, 175
Henry, Elizabeth: *Orpheus with His Lute,* 20-21
  'What Is a Poet?' 20
hermaphrodite, *136,* 141
Hermes, 69, 71, 73-80, 152. *See also* Mercurius/Hermes
Hesiod, 10, 115
hexagon/hexagonal, 12-13, 19, 90, 102, 119, 137
hieroglyphics, 68, *70*
Hierosgamos, 15n, 32, 114, 137
  alchemical, 56, 141-144
Hillary, Sir Edmund, 110
Hinduism, 51, 66n, 73, 105-109
hive(s), *passim,* 18-23, 46, 104, 115-116, 130, 169, *177*
  and altruism, 19, 21, 37, 40, 134n, 135
  and balance in nature, 46
  brood box in, 12, 13
  circumambulation of, *176. See also* circumambulation of the centre
  dream of, 166-169
  in Freemasonry, 102-104
  Mary and Christ painted on, *97*
  messages of, 101-102
  no second queen in, 31
  as prototype of first human society, 51
  renewal of, 139-141. *See also* swarm/swarming
  self-regulation of, 9, 21, 30-31, 33, 37, 46, 87, 141

task differentiation in, 19-20, 30, 135
'unity in multiplicity', 21, 135, 159
well-regulated, 30, 39-41
'holy rage', 117
homeopathy, 22, 119
Homer (Homeric): 64
Hymn to Dionysos, 56-57
Hymn to Hermes, 73-76
Hymn to the Muses and Apollo, 64
The Illiad, 15n
honey, *passim,*
Colony Collapse Disorder, 13n
'cured', 26
in death rituals, 88
as elixir of life, 57, 130, 143, 159
as embalming fluid, 88, 91
and ethics, 95
as fecundating principle, 107, 159
in Fifth Brahmana, 107
flow, 13-14, 171
as food of the gods, 9, 23, 46, 50, 58, 76-77, 87-88, 96
as funeral food, 88
and healing, 9, 11, 14, 20-25, 33, 59-60, 67, 71-72, 75, 78-79, 88-89, 96, 105, 110, 115, 118-122, 143
-hunters, rock art, *111*
'I Turn Everything to Honey', *99*
in marriage rituals, 88-90
medicinal use, 22, 91, 104-105
negative aspect, 90
never putrifies, 88
and Orpheus cult, 58
as purest food, 50-51
'The Rose Gives Honey to the Bee', *102*
in Sanskrit, 105
and sensuality, 57, 131
as symbol of deep eros and love, 11, 57, 121, 124, 131
unstrained, 13
See also Transformed Berserker
honeycomb, 9, 12-14, 19, 25-26, 90, 96, 153, 167, 169
Horae, 60
housework, 118, 127
Hughes, Ted, 'The Bee God', 84, 114
Hymen, 65
hymenopteran order, 65

Ibn Magih, 104
incarnation, 100
incubation, 25n, 78
rituals, 72, 89
individuation process, 10, 24, 68, 73n, 77, 91, 95, 114-116, 128, 130-169
Indra, 105
initiation rituals, 50, 72, 78, 87, 89, 93, 127
Inanna, 113
inner order, archetype of, 9-10, 31-37, 40, 49, 65, 94, 101, 114, 118, 127, 151, 157, 175. *See also* mandala
insect socio-biology, 45
inspiration, 20, 27, 31, 64-71, 73-77, 124
instinct(s), instinctive/instinctual, 10, 14, 17, 21, 41-43, 60, 112, 118, 124, 127, 130, 138, 153
interregnum, 32-36, 127, 139. *See also* swarm/swarming
introversion, 122, 131-132, 158
intuition/intuitive, 77-78, 142, 155
Islam, 104-105

Jacob's ladder, 79
John of the Cross, St: 'Spiritual Canticle', 100
Judeua, Philo, 13n
Judaism, 95-96
Jung/Jungian, C. G., *passim,*
analysis, 23
on archetypes, 41-42
and colour spectrum, 41
dream of ship coming to port, 117
on Eros, 118
on grounding himself, 117
on *hierosgamos,* 114-115
on love, 108
and power, 121
on mandala symbolism, 151-152, 156-157
Memories, Dreams, Reflections, 78, 117
Mysterium Coniunctionis, 78, 131, 137, 144, 155, 161
on the number 6, 13n
on 'objective knowledge', 108
Psychological Types, 117
on the Self, 10, 21, 66, 133
on sexuality, 113-115
on spirit and instinct, 118
'Synchronicity: an Acausal Connecting Principle', 24, 43
and tapas, 25
'The Tavistock Lectures', 97
on Wotan, 125, 128n
justice, 115

ka soul, 94
kabir, 78
Kalahari folk tale, 80
Kalisasa, 105
Kama, 105, *106*
Kama Sutra, 105
Karpo, 60
Kephri, 95n
Kerényi, Karl: *Dionysos,* 10n, 46n, 55-56, 63, 141
The Heroes of the Greeks, 58
Kiaochau, Rainmaker of, 122
killer bees, 33-34
king bee, 68-71
King Solomon, 11, 144-145, 144n
Komarios to Cleopatra, 14, 25-26, 140n, 142, 149, 155
Koran, 104-105
Krishna, 105
Kundalini, 26-27
Kyrene, 58, 60, 62

Langstroth: *On the Hive and the Honeybee,* 14n
lapis, 9, 15n, 24, 33, 68, 80, 115, 128-133, 136-137, 142, 146-148, 154-156, 158, 163, 165, 173
and *multiplicatio,* 133-134, 148
laziness, 10, 19, 90, 116-117, 131
lion, 84, *85,* 96
logos, 73, 78, 147
love, 100, 108, 112-113, 118-119, 122-125, 146, 170. *See also* eros
and power, 121-122
and war, 10, 115-119, 128
lumen naturae, 68
lunar consciousness, 27, 73n, 77-78, 118
Luton, Frith, 12-16, 38n, 121, 150, 160, 173-175
her *periginatio animae,* 121

maenads, 56, 74
madharva, 105
madheri, 105

madhu, 105, 107-108
Maeterlinck, Maurice: *The Life of the Bee,* 84, 86, 138, 140-141
Maia, 75-77
Maier, Michael, 118, 173
mandala symbolism, 10, 15n, 101, 151-157. *See also* inner order, archetype of
Mary mother of God, *97-98,* 100
maternal, 33n, 91, 144n
matrifocal, 47
mead, 59, 63, 88-89
meditation, 109
Medusa, 48
Meier, C. A: *Healing Dreams and Ritual,* 72
'The Psychological Parameters of the Examination Situation', 34
melissa/melissae, 27, 51, 65-67, 72
Mercurius/Hermes, 55, 58, 73-80, 83-84, 124, 131, 136, 140, 141, 144, 145, 148, 162, 163
Mercury, 78-79, 145, 148
Merope, 48
Michelet, Jules, 53
Mills, John: 'An Essay on the Management of Bees', 22
Minoans, 47-50, 89
Mithras, 89
Mohammed, 104-105
mother complex, 90n, 117n
Mount Everest, 110
multiplicatio, 63-64, 68, 122, 133-134, 146-150, *162, 164,* 167-8
and eros, 162-166
and flowering, 146-151
image from *Alchymie, 162*
image from *Cabala, 164*
Muses and divination, 10, 58-60, 64-67
mystery of the bees, 9, 40, 45-46, 135
'Mystery of the Queen', 45, 139, 141, 144-146, 153

natural healing, 22-23
navigational skills, 27-28, *29-30,* 44-45, 158. *See also* dance language
nectar, 9-11, 14, 16-17, 19, 22-28, 30, 42-43, 50, 56, 58, 87-88, 91, 105-106, 110, 116, 129, 151, 172
nectar, honey and tapas, 14, 17, 19, 24-26, 42
Nefertem, 81
Neumann, Erich: *The Origins and History of Consciousness,* 68-69
nigredo, 31-32, 75, 116, 131, 142, 154
Nizami, 42, 113
Norgay, Tensing, 110
number 6, 13n
numinosity, 9, 15, 18, 39, 45-46, 66-69, 80, 86-87, 96, 135, 138, 145, 151, 175
nuptial flight of queen bee, 18, 32 137-139, 140, 147
Nut, 91, *92*

'objective knowledge', 108
omphalos as world navel, 65-66, *67*
'Opening of the Mouth' ceremony, *91,* 93
opposites, 10, 31, 36-37, 43, 50, 90, 112-119, 128-130, 131, 134, 137, 141, 149, 153, 171
opus, 63, 130, 133, 146, 155, 163
oracles, 65, 71
order. *See* inner order, archetype of

'orphan', 15n
Orpheus, 57-58, 60
Ovid, 53, 56
ox-herding pictures, 133

Paracelsus, 90
Paradise, 78
parthenogenesis, 18, 89. *See also* bugonia
Paschal candle, 97
passion, 115-117, 121, 124, 128, 146-147, 166
Pausanias: *The Description of Greece,* 71-72
Penotus, 22, 144
peregrinatio animae (pilgrimage*),* 126
Persephone, 51
philosopher's stone, *see* lapis
Philostratus the Elder, 66
pilgrim/pilgrimage, 12, 74, 121-124, 126-128, *160*
Pindar, 64-65
  Pythian, 58
Plath, Sylvia: 'Stings', 114
  'Wintering', 32
Plato, 64
Pliny, 79
  poetry, 20, 58, 67, 114
pollen, 13-16, 19, 22-25, 27-28, 30, 43n, 89, 110, 172
  pollen path ceremonies, 89
pollination, 11, 13, 14, 36-38, 46, 50, 80, *99, 102,* 130, 131, 146
  Porphyry, 62, 65
  Poseidon, 62
  Post-Mesolithic rock art, *111*
  prima materia, 24, 33n, 148, 154, 155
  projectio, 63, 163
  projections, 108
propolis, 19, 23, 25
Proteus, 62, 131
psychological/psychology, *passim,* 13-14, 17, 24, 29, 31, 34, 66, 73n, 77, 94, 100, 107-108, 112-132, 152-154, 157, 165, 172
  crisis, 117, 124, 127
psychopomp(s), 14, 77-79, 81, 128
psychosis, 33
puer aeternus, 90n
Purchas, Samuel, 21
Purusha, 108
pyramid building, 69n
Pythagoras, 64
Pythia, 66

quadratura circuli, 156
queen bee, 12-13, 18-23, 65, 68, 74, 87, 104, 136-138, 140-141, 146, 149
  killing of, 31
  and king bee, 68-71
  'Mystery of the Queen', 144-146, 153
  retinue of, 33, 151
  no second queen, 31, 33
  nuptial flight of, 18, 32, 51, 56, 137-139, 140-141, 147
  as symbol of the Self, 33
Queen of Sheba, 95, 144-145
quintessence, 137, 155-156, 159, 160

rain, as communication from the gods, 79
rainmaker/rainmaking, 59, 71-73, 79, 133, 147
  of Kiaochau, 122
Ransome, Hilda: *Sacred Bee,* 93
Re, 69, 80, 92-93

Re's tears, 68, 91-94
redemption, 129
religious associations, 86-111
Rig-Veda, 88, 107-108
rites of passage, 34, 87-88. *See also* initiation rituals
robber bees, 19, 37
Roberts, Allen: *The Craft and Its Symbols,* 103-104
Robinson, Gene, 38, 44-45
rotation. *See* circulatio
royal jelly, 19, 23, 51, 167n
rubedo, 131, 142, 153, 154
Rumanian folk tale, 81-83

St Niklausen, 12, 26, 168, 174
Salt Magical Papyrus, 94
Samson, 58, 64, 95-96, 147
  and Delilah, 90, 95-96
Sappho, 64
Satapatha Brahmana, 88
Schweizer, Andreas, 36, 140n, 154, 163, 165
scintillae, 75, 165
scout bees, 27-28, 32, 43
Self, 10, 12, 20-21, 33, 37, 45, 68-69, 73n, 75, 81, 101, 107-108, 122, 126, 130, 132, 133, 146, 149, 151
self-regulation of the hive, *passim,* 9, 21, 30-31, 33, 37, 46. *See also* circumambulating the centre
senarious (6), 13n
Seneca, 41
separation, 34
sex/sexuality, 113-114
shadow, 31, 78, 100, 121, 127-128, 171
shamanism, 56n, 78, 124, 131, 147
solar consciousness. *See* logos
solificatio, 32, 145
Soma, 106
soul, 69, 114, 129, 144n
Spes, 27, *52*
spirit, ascending and descending, 56, 72-73, 124, 141, 142
  chthonic, 113
  divine, 64, 75
  and instinct, 118
  and matter, 124
  of truth, 124
spiritual pilgrim, 159, *160*
spontaneous regeneration of life. *See* bugonia
squaring, of circle, 151, 156, 175. *See also* quadratura circuli
Srinivasan, Mandyam, 44
Steiner, Rudolf, 113, 135, 146, 149-150
sting/stinging, 9, 13, 22, 33, 46, 80, 83, 109, 114-115, 118-119, *120,* 126, 130, 131, 169, 174
  as source of healing, 22
stone, 66, 78, 173, 174. *See also* lapis
'stone the builders rejected', 15
Swammerdam, Jan: *The Book of Nature* – drawings of bee anatomy, *120*
swarm/swarming, 32-34, *35,* 36, 74-76, 97, 100-101, 127, 131, 139-141
sweetness, 8, 9, 10, 26, 42, 57, 74, 83, 87-88, 90, 94-96, 100, 106, 110, 113, 118-119, 143, 147-148, 155n, 159, 173
symbolism in myth, legend and story, 10, 46-85
synchronicity, 24, 27, 44, 163
tao, 39, 73n, 131, 132, 163
tapas, 24-25

Taurus, 65, 84, 148
Telesphoros, 78
'telling of the bees', 39
temenos, 23, 78
Thallo, 60
Theocritus, 'The Robber of Honey-combs', 119-120
theos nomios, 59
'Thieves', 55-56
three, 75
Thriae, 74-75
Titans, 60, 74
Tolstoy, Leo, 100-101
'The Kingdom of God Is Within You', 101
War and Peace, 171-172
transcendence/transcendent, 155-159, *160*
transcerebral thought and perception, 44
transformation(s), 50, 53, 55-56, 63, 77-79, 95n, 96-97, 100-101, 122, 124-127, 130-169, 171
Transformed Berserker, 11-12, 38, 46, 107n, 108, 112-113, 117, 121-129, 133, 142, 143, 148, 169, 171, 174
Triple Goddess, 64
Trophonius, rainmaker, *71,* 72-73, 89
types of bees, *17,*18-45. *See also* drones; queen bee; workers

ultra-violet and infra-red, 41-45
'unalloyed man', 127
unconscious, descent into, 117
unio corporialis, 158
unio mentalis, 155, 157, 158
'unity in multiplicity', 21. 135, 159
University of Western Sydney, 166-168
unus mundus, 137, 148, 158-159, *160,* 163
Upanishad, 107
Uranus, 58, 60
urticaria, 22

Varro, 79
Varroa destructor mite, 167n
Venus, 65, 84, *85,* 90, 113, 119, 148, 174
Vestal virgins, 51
Virgil, 18, 20-21, 79
Aeneid, 116
Georgics, 38, 53, *54,* 59, 62, 86
Virgin Mary. *See* Mary mother of God
virgin queens, 31, 51
virginity, 51, 97, 114
Virgo, 114-115
Van Over, Raymond: *Sacred Texts,* 107-108
Von Flüe, Niklaus, 11-12, 38, 121-129, 171
life-changing vision of, 122-129, 133, 142, 148, 171, 174
Von Franz, Marie-Louise, 11-12, 17, 131-132, 142, 147, 150, 173-5
Alchemy, 116
Archetypal Dimensions of the Psyche, 109, 122-124
The Cat, 34
The Feminine in Fairy Tales, 19n, 117n,
On Dreams and Death, 14n-15n
Projection and Re-Collection in Jungian Psychology, 170
The Passion of Perpetua, 89
Shadow and Evil in Fairy Tales, 17, 43, 90n

Von Franz *(cont.):*
'Transformed Berserker' lecture, 11-12, 38, 46, 107n, 108, 121-129, 148, 171
Von Frisch, Karl, 8, 24, 27-28, *29-30,* 38, 41-45
The Dance Language of Bees, 45
Von Gaertringen, Hiller: *Aristaios,* 59

war, and love, 113-118, 174
wasps, 41-42
water(s), 14, 15, 19, 46, 57, 62, 79, 80, 82, 89, 90, 94, 97, 107, 108, 130, 131, 141-144, 147, 150, 161, 173, 174
wax. *See* beeswax
weeds, 15-16
'white foliated earth', 100
Wilson, Bee: *The Hive,* 34, 38, 88, 101
wintering, 31-32, 131
'women washing', 116
work, and inertia, 31
worker bees, *17,* 18-19, 21, 26-30, 33, 50, 84, 113
Wotan, 124-126, 128, 131
wounding and healing, 20, 119-121

Yeats, William Butler: 'The Second Coming', 36
yellowing, 162-166
yoga, 27
Yuthok, Lama Choedak: *Healing Relationships,* 109

Zeltzer, Charles, 27
Zen Buddhism, 118, 133
Zeus, 51, 57, 64, 66, 75-79
birth of, 55-57, 64, 129
and Hera, 15n
Icmaios, 59
zoë, 45, 58, 87, 130, 149
zoology, 65

Also in this Series, by Daryl Sharp

Please see last page for discounts and postage/handling.

THE SECRET RAVEN
Conflict and Transformation in the Life of Franz Kafka
ISBN 978-0-919123-00-7. (1980) 128 pp. $25

PERSONALITY TYPES: Jung's Model of Typology
ISBN 978-0-919123-30-9. (1987) 128 pp. **Diagrams** $25

THE SURVIVAL PAPERS: Anatomy of a Midlife Crisis
ISBN 978-0-919123-34-2. (1988) 160 pp. $25

DEAR GLADYS: The Survival Papers, Book 2
ISBN 978-0-919123-36-6. (1989) 144 pp. $25

JUNG LEXICON: A Primer of Terms and Concepts
ISBN 978-0-919123-48-9. (1991) 160 pp. **Diagrams** $25

GETTING TO KNOW YOU: The Inside Out of Relationship
ISBN 978-0-919123-56-4. (1992) 128 pp. $25

THE BRILLIG TRILOGY:

1. CHICKEN LITTLE: The Inside Story ***(A Jungian romance)***
ISBN 978-0-919123-62-5. (1993) 128 pp. $25

2. WHO AM I, REALLY? Personality, Soul and Individuation
ISBN 978-0-919123-68-7. (1995) 144 pp. $25

3. LIVING JUNG: The Good and the Better
ISBN 978-0-919123-73-1. (1996) 128 pp. $25

JUNGIAN PSYCHOLOGY UNPLUGGED: My Life as an Elephant
ISBN 978-0-919123-81-6. (1998) 160 pp. $25

DIGESTING JUNG: Food for the Journey
ISBN 978-0-919123-96-0. (2001) 128 pp. $25

JUNG UNCORKED: Rare Vintages from the Cellar of Analytical Psychology
Four Books. ISBN 978-1-894574-21-1/22-8.. (2008) 128 pp. each. $25 each

THE SLEEPNOT TRILOGY:

1. NOT THE BIG SLEEP: On having fun, seriously ***(A Jungian romance)***
ISBN 978-0-894574-13-6. (2005) 128 pp. $25

2. ON STAYING AWAKE: Getting Older and Bolder *(Another Jungian romance)*
ISBN 978-0-894574-16-7. (2006) 144 pp. $25

3. EYES WIDE OPEN: Late Thoughts *(Another Jungian romance)*
ISBN 978-0-894574-18-1.. (2007) 160 pp. $25

Also in this Series, by Edward F. Edinger

SCIENCE OF THE SOUL: A Jungian Perspective
ISBN 978-1-894574-03-6. (2002) 128 pp. $25

THE PSYCHE ON STAGE
Individuation Motifs in Shakespeare and Sophocles
ISBN 978-0-919123-94-6. (2001) 96 pp. **Illustrated** $25

EGO AND SELF: The Old Testament Prophets
ISBN 978-0-919123-91-5. (2000) 160 pp. $25

THE PSYCHE IN ANTIQUITY
Book 1: Early Greek Philosophy
ISBN 978-0-919123-86-1. (1999) 128 pp. $25
Book 2: Gnosticism and Early Christianity
ISBN 978-0-919123-87-8. (1999) 160 pp. $25

THE AION LECTURES: Exploring the Self in Jung's *Aion*
ISBN 978-0-919123-72-4. (1996) 208 pp. **30 illustrations** $30

MELVILLE'S MOBY-DICK: An American Nekyia
ISBN 978-0-919123-70-0. (1995) 160 pp. $25

THE MYSTERIUM LECTURES
A Journey Through Jung's *Mysterium Coniunctionis*
ISBN 978-0-919123-66-3. (1995) 352 pp. **90 illustrations** $40

THE MYSTERY OF THE CONIUNCTIO
Alchemical Image of Individuation
ISBN 978-0-919123-67-6. (1994) 112 pp. **48 illustrations** $25

GOETHE'S FAUST: Notes for a Jungian Commentary
ISBN 978-0-919123-44-1. (1990) 112 pp. $25

THE CHRISTIAN ARCHETYPE A Jungian Commentary on the Life of Christ
ISBN 978-0-919123-27-4. (1987) 144 pp. **34 illustrations** $25

THE BIBLE AND THE PSYCHE
Individuation Symbolism in the Old Testament
ISBN 978-0-919123-23-1. (1986) 176 pp. $30

ENCOUNTER WITH THE SELF
A Jungian Commentary on William Blake's *Illustrations of the Book of Job*
ISBN 978-0-919123-21-2. (1986) 80 pp. **22 illustrations** $25

THE CREATION OF CONSCIOUSNESS
Jung's Myth for Modern Man
ISBN 978-0-919123-13-7. (1984) 128 pp. **10 illustrations** $25

Also in this Series, by Marie-Louise von Franz

AURORA CONSURGENS: On the Problem of Opposites in Alchemy
ISBN 978-0-919123-90-8. (2000) 576pp. **30-page Index** ***Sewn*** $50
A penetrating commentary on a rare medieval treatise, scattered throughout with insights relevant to the process of individuation in modern men and women.

THE PROBLEM OF THE PUER AETERNUS
ISBN 978-0-919123-88-5. (2000) 288pp. **11 illustrations** ***Sewn*** $40
The term *puer aeternus* (Latin, eternal youth) is used in Jungian psychology to describe a certain type of man or woman: charming, creative, and ever in pursuit of their dreams. This is the classic study of those who remain adolescent well into their adult years.

THE CAT: A Tale of Feminine Redemption
ISBN 978-0-919123-84-7. (1999) 128pp. **8 illustrations** ***Sewn*** $25
"The Cat" is a Romanian fairy tale about a princess who at the age of seventeen is bewitched—turned into a cat. . . . One by one von Franz unravels the symbolic threads.

C.G. JUNG: His Myth in Our Time
ISBN 978-0-919123-78-6. (1998) 368pp. **30-page Index** *Sewn* $40
The most authoritative biography of Jung, comprising an historical account of his seminal ideas, including his views on the collective unconscious, archetypes and complexes, typology, creativity, active imagination and individuation.

ARCHETYPAL PATTERNS IN FAIRY TALES
ISBN 978-0-919123-77-9. (1997) 192pp. *Sewn* $30
In-depth studies of six fairy tales—from Spain, Denmark, China, France and Africa, and one from the Grimm collection—with references to parallel themes in many others.

REDEMPTION MOTIFS IN FAIRY TALES
ISBN 978-0-919123-01-4. (1980) 128pp. *Sewn* $25
A nonlinear approach to the significance of fairy tales for an understanding of the process of psychological development. Concise explanations of complexes, projection, archetypes and active imagination. A modern classic.

ON DIVINATION AND SYNCHRONICITY
The Psychology of Meaningful Chance
ISBN 978-0-919123-02-1. (1980) 128pp. **15 illustrations** *Sewn* $25
A penetrating study of the psychological aspects of time, number and methods of divining fate such as the I Ching, astrology, Tarot, palmistry, dice, etc. Extends and amplifies Jung's work on synchronicity, contrasting Western attitudes with those of the East.

ALCHEMY: An Introduction to the Symbolism and the Psychology
ISBN 978-0-919123-04-5. (1980) 288pp. **84 illustrations** *Sewn* $40

Designed as an introduction to Jung's weightier writings on alchemy. Invaluable for interpreting images in modern dreams and for an understanding of relationships. Rich in insights from analytic experience.

Also in this Series, by James Hollis

THE MIDDLE PASSAGE: From Misery to Meaning in Midlife
ISBN 0-919123-60-0. (1993) 128pp. ***Sewn*** $25
Why do so many go through so much disruption in their middle years? Why then? What does it mean and how can we survive it? Hollis shows how we can pass through midlife consciously, rendering the second half of life all the richer and more meaningful.

UNDER SATURN'S SHADOW: The Wounding and Healing of Men
ISBN 0-919123-64-3. (1994) 144pp. ***Sewn*** $25
Saturn was the Roman god who ate his children to stop them from usurping his power. Men have been psychologically and spiritually wounded by this legacy. Hollis offers a new perspective on the secrets men carry in their hearts, and how they may be healed.

TRACKING THE GODS: The Place of Myth in Modern Life
ISBN 0-919123-69-4. (1995) 160pp. ***Sewn*** $25
Whatever our religious background or personal psychology, a greater intimacy with myth provides a vital link with meaning. Here Hollis explains why a connection with our mythic roots is crucial for us as individuals and as responsible citizens.

SWAMPLANDS OF THE SOUL: New Life in Dismal Places
ISBN 0-919123-74-0. (1996) 160pp. *Sewn* $25
Much of our time on earth we are lost in the quicksands of guilt, anxiety, betrayal, grief, doubt, loss, loneliness, despair, anger, obsessions, addictions, depression and the like. Perhaps the goal of life is not happiness but meaning. Hollis illuminates the way.

THE EDEN PROJECT: In Search of the Magical Other
ISBN 0-919123-80-5. (1998) 160pp. *Sewn* $25
A timely and thought-provoking corrective to the fantasies about relationships that permeate Western culture. Here is a challenge to greater personal responsibility—a call for individual growth as opposed to seeking rescue from others.

CREATING A LIFE: Finding Your Individual Path
ISBN 0-919123-93-7. (2001) 160pp. *Sewn* $25
With insight and compassion grounded in the humanist side of analytical psychology, Hollis elucidates the circuitous path of individuation, illustrating how we may come to understand our life choices and relationships by exploring our core complexes.

ON THIS JOURNEY WE CALL OUR LIFE: Living the Questions
ISBN 1-894574-04-4. (2003) 160pp. *Sewn* $25
This book seeks a working partnership with readers. Hollis shares his personal experience only so that we may more deeply understand our own. It is a partnership rich in poetry as well as prose, but most of all it reminds us of the treasures of uncertainty.